3D Printing with Fusion 360

Design for additive manufacturing, and level up your simulation and print preparation skills

Sualp Ozel

BIRMINGHAM—MUMBAI

3D Printing with Fusion 360

Group Product Manager: Rohit Rajkumar

Publishing Product Manager: Vaideeshwari Roshan

Senior Editor: Aamir Ahmed

Technical Reviewer: Chetan Gowda HS

Book Project Manager: Aishwarya Mohan

Technical Editor: Simran H. Udasi

Copy Editor: Safis Editing

Proofreader: Safis Editing

Indexer: Rekha Nair

Production Designer: Prafulla Nikalje

DevRel Marketing Coordinator: Anamika Singh

First published: November 2023

Production reference: 2031225

Published by Packt Publishing Ltd.
Grosvenor House
11 St Paul's Square
Birmingham
B3 1RB, UK

ISBN 978-1-80324-664-2

www.packtpub.com

I dedicate this book to my family. I love you with all my heart. To my wife, Marlene, you have been my champion and biggest supporter, not only while authoring this book, but throughout our joint life journey. To my sons, Kerem and Koray, thank you for listening to me during countless dinner conversations about 3D printing and for all your ideas about what to 3D print next. To my parents, Bijen and Suat. Thank you for your endless support, love, and inspiration.

Foreword

I have known and worked with Sualp Ozel for close to a decade, as one of the leading experts in Design Simulation and Additive manufacturing, he is my *go to* whenever I need to understand new developments and applications in the space. In this book, Sualp takes you on a learning journey from Design for Additive manufacturing, structural simulation, and print preparation through to process simulation and workflow automation.

The first section will get you familiar with the basics of Autodesk Fusion, from how to open new files or upload your existing models to the Fusion software for storage. Sualp demonstrates how to use Fusion to inspect and repair broken mesh files along with how to edit models so that they can be 3D printed effectively using various 3D-printing technologies. The section also covers how to use Fusion's Automated Modeling, Topology Optimization and Stress Simulation capabilities in tandem to generate functional parts that meet performance and manufacturing criteria.

The book also covers hollowing and latticing parts and the various ways you can create lightweight and material saving models along with how to export them to common slicers such as Cura, PreForm and Prusa Slicer.

In the third section you learn all about the *Manufacture* workspace of Fusion, how to utilize it as a native slicer to prepare your parts for 3D printing. From how to use Fusion effectively to select a printer and print settings, arrange and orient your parts within your build volume, create support structures based on the needs of your specific printer.

The final section is all about metal printing and automation, where Sualp draws from his years of experience and expertise to explain the unique requirements of metal powder bed fusion printing, how to create efficient support structures and simulate the printing process to detect potential build failures and compensate for thermal distortion.

Understanding thesD-pritical areas can save hours of engineering time and the expense of failed prints. Throughout this book, Sualp provides all you need to learn the entire workflow from design to 3D print using the powerful capabilities in Fusion in clear and easy to read resource suitable for people just starting their 3d printing journey, through to those that are ready to adopt some of the most advanced design and simulation tools available.

Duann Scott

3MF Consortium Executive director

Teaching Assistant, MIT xPRO

Founder at Bits to Atoms and the CDFAM Computational Design Symposium Series

Contributors

About the author

Sualp Ozel is a professional engineer and a senior product manager at Autodesk Inc. Sualp has 15 years of experience in planning roadmaps based on market requirements to deliver timely enhancements to existing products and go to market with new offerings while meeting the market's expectations and requirements for quality. Since 2016, he has managed Autodesk's additive manufacturing software portfolio including Fusion, Netfabb, Netfabb Local Simulation, and Within Medical. Sualp has also been an adjunct instructor at Carnegie Mellon University since 2014, teaching the *Introduction to CAD and CAE tools* course within the mechanical engineering department.

I want to thank everyone who supported me in this journey.

Thank you to Anthony Graves for making sure I had the best HP Z2 workstation so I could work with Autodesk Fusion seamlessly.

Thank you to Pavel Pelčák and the Prusa research organization for providing me with access to FFF and SLA 3D printers and materials to use when writing the book.

I also want to thank everyone who worked on Autodesk Fusion. Without their work and commitment, the design-and-make community would not be where it is today.

Table of Contents

Part 1: Design for Additive Manufacturing (DFAM) and Fusion

1

2

Part 2: Print Preparation – Creating an Additive Setup

6

Introducing the Manufacture Workspace for Print Preparation   161

7

Creating Your First Additive Setup   189

Part 3: Print Preparation – Positioning Parts, Generating Supports, and Toolpaths

8

Arranging and Orienting Components   219

Part 4: Metal Printing, Process Simulation, and Automation

Preface

As 3D printing is becoming mainstream, the demand for **Computer-Aided Design (CAD)** users with manufacturing knowledge is growing. You may have noticed that new capabilities around automated modeling, generative design, and additive manufacturing are now available in Autodesk Fusion. If you are interested in learning how you can benefit from these tools and improve your design and 3D-printing skills using Fusion, this book is for you.

In this book, you will learn how to open CAD and mesh files in Fusion, repair and edit them, and prepare them for 3D printing. In the context of print preparation, you will be introduced to print settings, support structures, and part orientation. The book will also highlight the various preferences of Fusion for additive manufacturing. In subsequent chapters, you will learn about choosing the right orientation and creating appropriate support structures based on printing technology and you will simulate the printing process to detect and remedy common print failures associated with the metal powder bed fusion process. This book will also cover how to arrange parts depending on the printing technology. By the end of this book, you will be acquainted with utilizing templates and scripts to automate common tasks around print preparation.

By the end of this book, you will have gained all the knowledge necessary to use Fusion for additive manufacturing.

> **Important note**
> This book has been authored in my own personal capacity and the views are my own, and not as an Autodesk employee, and the information in this book does not necessarily represent the views of Autodesk or its partners.

Who this book is for

If you're a designer using Autodesk Fusion daily and would like to learn how to 3D print your designs, or if you're interested in creating functional yet lightweight prints, this book is for you. Intermediate-level users of Fusion will benefit from having this book as a reference for **design for additive manufacturing (DFAM)** and print preparation.

Ideally, you need a rudimentary understanding of the design capabilities of Fusion before you start reading this book. Being able to create basic designs and knowing how to open existing CAD or mesh files with Fusion will allow you to get the most out of this book.

What this book covers

Chapter 1, *Opening, Inspecting, and Repairing CAD and Mesh Files*, explains how, in order to 3D print an object, you need a 3D model or the sliced version of that model, typically referred to as a G-code file. You don't always have to design the object in CAD software yourself. There are various websites where you can download solid models or mesh files. To go from design to 3D printing, you generally need a slicing software. But before you can slice a model you downloaded from the web, it is a good idea to make sure that the model is free of errors and is the correct size for your printer. In this chapter, we will open CAD and mesh files from third-party sources using Autodesk Fusion, check them for errors, and repair them if necessary.

Chapter 2, *Editing CAD/Mesh Files with DFAM Principles in Mind*, covers how Fusion can be used with or without design history being captured. In this chapter, we will highlight how to enable capturing design history after opening non-native CAD files. You will be introduced to editing CAD files you open in Fusion based on a set of DFAM principles. This chapter will also explore how to edit mesh files directly. The chapter will conclude by showcasing how to convert mesh files into solid objects to be edited for manufacturing.

Chapter 3, *Creating Lightweight Parts, and Identifying and Fixing Potential Failures with Simulation*, discusses one of the main benefits of additive manufacturing, which is the ability to create lightweight parts. The automated modeling capabilities within the *Design* workspace and the shape optimization analysis type within the *Simulation* workspace of Autodesk Fusion are great tools for designers to create organic-looking lightweight components. In this chapter, we will showcase how to create such components. We will also simulate them using linear static stress in order to detect possible failures our 3D-printed designs may experience and remedy such issues with targeted modeling modifications.

Chapter 4, *Hollowing and Latticing Parts to Reduce Material and Energy Usage*, explains how, when manufacturing large-volume components using **Fused Filament Fabrication** (**FFF**) 3D printers, we can make them strong yet lightweight by utilizing a combination of multiple contour lines around the perimeter in combination with a sparse infill. However, this same technique does not work for other 3D-printing technologies such as stereolithography, multi-jet fusion, or selective laser sintering. When creating parts to prototype using those technologies, designers must proactively think about how to make lightweight parts to reduce material and energy usage. One method a designer can utilize is to hollow and lattice parts prior to printing to mimic the effects created using a contour and an infill. Designers also must include drain holes as well as plugs for the resin or the powder to escape during the post-processing phase. In this chapter, we will highlight various tools Fusion has to create such designs.

Chapter 5, *Tessellating Models and Exporting Mesh Files to Third-Party Slicers*, explains how Autodesk Fusion has a built-in slicer (which will be covered in the following chapters); however, Fusion's slicer does not support all the 3D printers on the market. There are certain 3D printers that have their proprietary slicing software. In certain organizations, a software manager may require a certain slicing software to be used by a manufacturing engineer. Or you may simply wish to use a slicing software you are already familiar with. In such cases, after going through the first four chapters of this book, you may want to export your models to your preferred slicer. In this chapter, we will go over the various methods with which you can export your Fusion models to your preferred slicer.

Chapter 6, Introducing the Manufacture Workspace for Print Preparation, explains how, when designers create an assembly of parts, they don't anticipate all the parts of that assembly to be 3D printed. Any part that is designated to be 3D printed may have to be altered before manufacturing it. Making changes to individual parts for 3D printing and keeping track of those changes can be tricky. Autodesk Fusion offers multiple methods to deal with such challenges. For users who want to manage the relationship between a design for assembly versus a design for manufacturing externally, the *derive* workflow is the ideal solution. For users who wish to manage this relationship in a single Fusion design, the *manufacturing model* is the way to go. In this chapter, we will go over how to utilize both of these workflows and highlight the benefits and drawbacks of each method, while showcasing common examples of design changes you may want to make for 3D printing.

Chapter 7, Creating Your First Additive Setup, describes how, in Autodesk Fusion, a critical first step for 3D printing is to create an additive manufacturing setup. This simple yet powerful dialog includes choices for selecting the 3D printer and the associated print settings. Understanding how the machine and print setting libraries operate and how to customize those libraries with your 3D printers and print settings will greatly reduce the time required for you to go from design to printed part. In this chapter, we will also highlight various Fusion preferences related to 3D printing, which will improve your experience and save you time.

Chapter 8, Arranging and Orienting Components, teaches you that regardless of whether you're printing a single component or hundreds of them at the same time, arranging them within the build volume and orienting them based on the desired outcome is an important consideration to get a successful print. This chapter will go over the various 3D-printing technologies and how to best arrange and orient components for each technology. In this chapter, we will highlight how to manually translate and orient components. We will also showcase how to automatically orient parts and arrange them based on a given criteria, such as minimizing the build height or finding an orientation that will result in a minimum support structure volume.

Chapter 9, Print Settings, explains how, once the printing orientation and position of a given component are decided, you can simply slice the model and generate the toolpath to send it to a 3D printer. In certain cases, it may be necessary to edit the default print setting to change a certain aspect of the printing process. For example, you may want to add a skirt or a brim to improve print bed adhesion. You may want to decrease the infill density or increase the number of contours. In some cases, you may want to use a different print setting for a different part of the print. All these possibilities and more will be explored in this chapter.

Chapter 10, Support Structures, teaches you that most mainstream additive manufacturing technologies require support structures to be generated and printed for sections of components that have overhanging surfaces. This chapter highlights the bar and volume support structures that can be used for both FFF and SLA/DLP printing and how they differ based on the printing technology. You will also learn about the base plate support type and how it can be used in conjunction with bar supports with pads for SLA/DLP printing.

Chapter 11, Slicing Models and Simulating the Toolpath, describes how, once all the components we wish to print are arranged, oriented, and properly supported, it is time to generate the additive toolpath. Then, we can visualize the slices layer by layer as well as the movement of the extruder and/or laser. Generating the additive toolpath is a one-click solution in Autodesk Fusion, but it is recommended to review the toolpath simulation to understand whether the print settings we have chosen will result in the desired outcome. This chapter will highlight how to create the additive toolpath and simulate it.

Chapter 12, 3D Printing with Metal Printers, explains how Fusion and its extensions can be utilized for creating additive setups for metal powder bed fusion machines. Fusion users who are interested in setting up their metal prints need to pay special attention to their part position and orientation based on certain machine parameters, such as recorter blade and gas flow directions. In addition, metal printing requires unique support structures in order to minimize material waste, decrease print time, and minimize post-processing effort. This chapter will touch on part orientation based on recorder blade and unique support structure settings for metal 3D printing.

Chapter 13, Simulating the MPBF Process, tells you that metal powder is an expensive form of 3D printing material. Therefore, any mistake or print failure is a costly one when it comes to metal 3D printing. Fusion users with Manufacturing extension access can simulate their metal powder bed fusion printing process to detect and rectify common print failure modes such as part distortion and recorter blade interference. In this chapter, we will highlight how to perform such analyses and make the necessary changes to avoid common print failure modes.

Chapter 14, Automating Repetitive Tasks, highlights how to automate various aspects of Fusion to minimize our interaction with the software to create an additive setup and generate a toolpath. If you plan on using Fusion for additive manufacturing regularly, this chapter is for you. We will start the chapter by highlighting existing automations within Fusion's ecosystem by introducing you to the Fusion App Store. Next, we will highlight how to customize Fusion's machine and print setting libraries to create a fully defined machine you can utilize when creating your setups. You will then learn how to customize your inputs for various operations, such as automatic orientation studies, part arrangement, and support structure generation. We will also demonstrate how to save those inputs as user defaults. Next, we will create and save templates, which combine multiple operations into a single file that can be reused on subsequent additive setups. We will end the chapter by highlighting how to utilize Fusion's programming capabilities by creating a Python script to automate the entire process of generating an additive setup, orienting parts, arranging them within the build plate, and simulating the additive toolpath.

To get the most out of this book

Software/hardware covered in the book	Operating system requirements
Autodesk Fusion	Windows and macOS
Autodesk Netfabb	
PrusaSlicer, Formlabs PreForm, and UltiMaker Cura	

Download the example project files

You can download the example project files for this book from GitHub at `https://github.com/PacktPublishing/3D-Printing-with-Fusion-360`. If there's an update to the code, it will be updated in the GitHub repository.

We also have other code bundles from our rich catalog of books and videos available at `https://github.com/PacktPublishing/`. Check them out!

Conventions used

There are a number of text conventions used throughout this book.

`Code in text`: Indicates code words in text, database table names, folder names, filenames, file extensions, pathnames, dummy URLs, user input, and Twitter handles. Here is an example: we can use the **Insert Mesh** command and insert the `Clip_broken.STL` file into our active Fusion design document.

Bold: Indicates a new term, an important word, or words that you see onscreen. For instance, words in menus or dialog boxes appear in **bold**. Here is an example: In addition to centering and grounding a mesh body, we can also use the in-canvas manipulators (arrows and rotation tools) or the numerical inputs section within the **INSERT MESH** dialog to reposition the mesh body until we are satisfied.

> **Tips or important notes**
> Appear like this.

Get in touch

Feedback from our readers is always welcome.

General feedback: If you have questions about any aspect of this book, email us at `customercare@packtpub.com` and mention the book title in the subject of your message.

Errata: Although we have taken every care to ensure the accuracy of our content, mistakes do happen. If you have found a mistake in this book, we would be grateful if you would report this to us. Please visit `www.packtpub.com/support/errata` and fill in the form.

Piracy: If you come across any illegal copies of our works in any form on the internet, we would be grateful if you would provide us with the location address or website name. Please contact us at `copyright@packt.com` with a link to the material.

If you are interested in becoming an author: If there is a topic that you have expertise in and you are interested in either writing or contributing to a book, please visit `authors.packtpub.com`.

Share Your Thoughts

Once you've read *3D Printing with Fusion 360*, we'd love to hear your thoughts! Scan the QR code below to go straight to the Amazon review page for this book and share your feedback.

https://packt.link/r/1-803-24664-2

Your review is important to us and the tech community and will help us make sure we're delivering excellent quality content.

Download a free PDF copy of this book

Thanks for purchasing this book!

Do you like to read on the go but are unable to carry your print books everywhere? Is your eBook purchase not compatible with the device of your choice?

Don't worry, now with every Packt book you get a DRM-free PDF version of that book at no cost.

Read anywhere, any place, on any device. Search, copy, and paste code from your favorite technical books directly into your application.

The perks don't stop there, you can get exclusive access to discounts, newsletters, and great free content in your inbox daily

Follow these simple steps to get the benefits:

1. Scan the QR code or visit the link below:

https://packt.link/free-ebook/9781803246642

2. Submit your proof of purchase.
3. That's it! We'll send your free PDF and other benefits to your email directly.

Part 1: Design for Additive Manufacturing (DFAM) and Fusion

In this part, we will show you how to open solid or mesh designs created in other software in Autodesk Fusion. You will learn how to create and edit models using parametric modeling and direct modeling techniques. You will learn how to work with mesh files to repair them both manually and automatically, using various methods within the software.

You will also learn how to create automated models as well as topology-optimized models to generate lightweight components for 3D printing. You will learn how to simulate models created by the software to make sure that they can perform as desired under operating conditions.

We will end the first part by going over how to hollow large-volume parts and introduce internal lattices to aid with additive manufacturing.

This part has the following chapters:

- *Chapter 1, Opening, Inspecting, and Repairing CAD and Mesh files*
- *Chapter 2, Editing CAD/Mesh Files with DFAM Principles in Mind*
- *Chapter 3, Creating Lightweight Parts, and Identifying and Fixing Potential Failures with Simulation*
- *Chapter 4, Hollowing and Latticing Parts to Reduce Material and Energy Usage*

1

Opening, Inspecting, and Repairing CAD and Mesh files

Welcome to this book, *3D Printing with Fusion 360*. Fusion is a great tool to create and edit designs using parametric or direct modeling methods. With the addition of automated modeling, generative design, and topology optimization technologies, Fusion users now have a plethora of methods to design for additive manufacturing. This book will provide the necessary steps for you to manufacture your designs with Fusion, using common 3D-printing technologies such as fused filament fabrication, stereolithography, selective laser sintering, binder jetting, and metal powder bed fusion. You will also get valuable tips and tricks on how to set up Fusion, which will allow you to get the most out of the software for your 3D-printing needs.

Fusion is a cloud-based 3D modeling, **Computer Aided Design (CAD)**, **Computer Aided Manufacturing (CAM)**, **computer aided engineering** (**CAE**), and **Printed Circuit Board** (**PCB**) design software platform for professional product design and manufacturing (`https://www.autodesk.com/products/fusion-360/overview`).

It is available for both Windows and macOS, with simplified applications available for both Android and iOS. Fusion 360's initial release dates back to 2013, which is around the same time that 3D printing became mainstream. During the 2010s, the maker community quickly adopted Fusion 360 for designing and manufacturing.

When manufacturing designs using 3D printing, mesh (STL) files were the only file types that early 3D printing slicer software would accept as input. As the maker community created and shared their designs publicly using STL files on web pages such as `thingiverse.com`, the number of 3D printable designs increased exponentially. Unfortunately, the STL file format has many problems and limitations. To fix those problems, the 3D printing community needed reliable and easy-to-use software.

Fusion always included functionality around working with both CAD and mesh files, making it a great tool for 3D printing. However, Fusion 360's mesh functionality was offered as a technology preview and needed to be turned on within Fusion 360's preferences. Autodesk – the parent company of Fusion 360 – releases new updates for Fusion 360 regularly. With the July 2021 update of Fusion 360,

Autodesk made significant changes to the mesh workflows, by graduating the mesh workspace from being a part of a tech preview to being a part of Fusion's Design workspace.

Today, you can use Fusion to open, inspect, and repair mesh files that you can download from numerous third-party sources.

In this chapter, we will start by looking at the various ways we can bring CAD/mesh data into Fusion by introducing you to Fusion Team and the Fusion user interface. We will go over how to create projects and folders to better organize our data. Next, we will show numerous methods to insert mesh files such as STL, OBJ, and 3MF into Fusion. We will end the chapter by inspecting the mesh data we insert into Fusion for potential defects and repairing those defects automatically and manually.

In this chapter, we will cover the following topics:

- Opening and uploading workflows for CAD models and mesh files
- Inserting Mesh workflows for STL, OBJ, and 3MF files
- Inspecting Mesh bodies and repairing them

By the end of this chapter, you will have learned how to open models created using other tools in Fusion. You will have learned how to insert mesh files into Fusion. You will also know how to inspect and repair mesh files using automatic and manual methods.

Technical requirements

Most of the topics covered in this book apply to all licensing levels (personal/hobby, start-up, educational, and commercial) of Fusion. However, the book will also cover advanced functionality not available to personal/hobby use licenses. Such advanced functionality will be identified as *N/A for Personal Use*, and *Requires Access to Extension* as appropriate.

Fusion has the following system-specific requirements:

- **Apple® macOS**:
  - macOS 13 Ventura – version 2.0.15289 or newer
  - macOS 12 Monterey
  - macOS 11 Big Sur
- **Microsoft Windows**:
  - Windows 11
  - Windows 10 (64-bit) – version 1809 or newer

- **CPU, memory, and disk space requirements**:

  - x86-based 64-bit processor (for example, Intel Core i and AMD Ryzen series), 4 cores, and 1.7 GHz or greater; 32-bit is not supported. Apple silicon processors require Rosetta 2.

  - 4 GB of RAM and 8.5 GB of storage.

> **Important note**
>
> A detailed list of system requirements can be found here: `https://knowledge.autodesk.com/support/fusion-360/learn-explore/caas/sfdcarticles/sfdcarticles/System-requirements-for-Autodesk-Fusion-360.html`.

The lesson files for this chapter can be found here: `https://github.com/PacktPublishing/3D-Printing-with-Fusion-360`.

Opening and Uploading workflows for CAD models and Mesh files

Fusion is a cloud-connected CAD/CAM/CAE and PCB design tool. You can create new designs or open designs created in other tools that are saved as CAD or mesh files. In this section, we will cover all the ways you can open models created by other software in Fusion so that we can start getting them ready for 3D printing.

As Fusion is cloud-connected software, it has cloud storage and data-sharing capabilities built in. As a Fusion user, you automatically get access to a cloud-connected data storage and management tool called Fusion Team. This tool also allows you to administer your team of Fusion users by giving you tools for permissions, version control, markup, and comments. According to an Autodesk Knowledge Network article, Fusion users automatically get unlimited storage for Fusion data and 500 GB of data storage for non-Fusion documents. That is a generous amount of storage to tackle any project.

> **Important note**
>
> For more information on how cloud data storage works, you can refer to this article at `https://knowledge.autodesk.com/support/fusion-360/learn-explore/caas/sfdcarticles/sfdcarticles/How-big-is-the-cloud-storage-for-Fusion.html`.

After creating an Autodesk account and gaining access to Fusion, regardless of your licensing level, you can log in to your Fusion Team account (`https://myhub.autodesk360.com/`) and start your Fusion data management journey.

If you are an individual user, you don't have to worry about changing teams. However, if you work for an organization with multiple Fusion users, your organization may have multiple teams already set up. For example, each user may have their own team. You can switch the active team by selecting the profile icon in the top-right corner of the screen, as shown in the following screenshot:

Figure 1.1 – My FUSION TEAM view with an OWNED BY ME filter applied

Figure 1.1 shows my Fusion TEAM view on a Google Chrome browser with the **OWNED BY ME** filter applied to the list of projects. If I were part of an organization with multiple teams, I would select the **Profile** icon in the top-right corner and change my active team to participate in a project stored in another team's hub. *Figure 1.1* also shows that, by default, we have two items that are owned by me – **Assets** and **My First Project**.

In *Figure 1.2*, you can see that by selecting **My First Project** and expanding **Project Details**, we can edit the project name and image and manage members who can contribute to and view the contents of this project:

Figure 1.2 – You can rename projects on the Project Details screen

Once the project has a name we approve (in *Figure 1.2*, I renamed the project 3D printing), we can take a look at various methods we have available to upload designs and other documents to Fusion Team and Fusion.

The first option we have to upload our designs and documents to Fusion Team is within the created project itself. As shown in *Figure 1.3*, we are in a project named **3D printing**, and we can select the blue **Upload** button to see the various options we have to upload things, such as **File**, **Folder**, **Assembly**, and **From Dropbox**. If we select **File** from that dropdown, we can upload any file stored on our local storage directly to our active project on **Fusion TEAM**. If the files we upload contain mesh or CAD geometry (for example, **STL**, **STEP**, **SAT**, **Inventor**, or **Solidworks**), we will be able to open those designs in Fusion for further editing and manufacturing. If they are non-CAD files (for example, Microsoft Word documents or PDF files), we won't have native access to them in Fusion, but we can still view them in our internet browser or mobile apps.

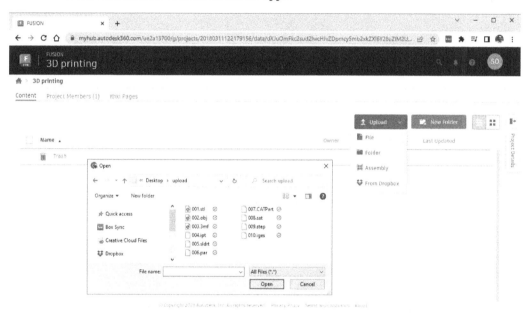

Figure 1.3 – Uploading local files to our 3D printing project on Fusion TEAM

If you are used to traditional desktop CAD software, uploading files to a Fusion project using FUSION TEAM may not have the intuitive workflow you are familiar with. However, it is a useful way to quickly upload multiple files associated with a project to your Autodesk cloud storage.

If you seek a more traditional workflow to open your CAD or mesh files, you will want to have Fusion installed on your computer. You can download and install Fusion from the following link: `https://www.autodesk.com/content/autodesk/global/en/products/fusion-360/trial-intake`.

Once Fusion is installed and started, you can access the **File** dropdown and select **Open**. When you do, you will be greeted with the **Open** dialog. As shown in *Figure 1.4*, select the **Open from my computer...** option within this dialog, and you will be able to select from a list of supported mesh and CAD file types, as shown in *Figure 1.5*:

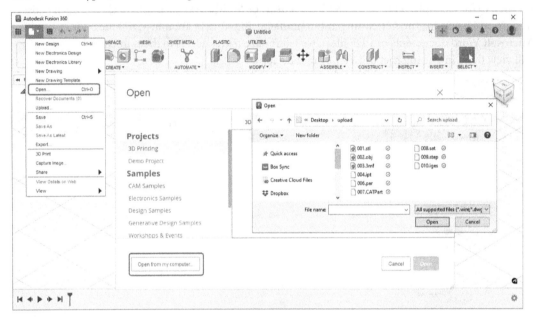

Figure 1.4 – The File | Open... workflow allows you to open CAD and mesh files from your computer

In this dialog, you will have the option to open any of the Mesh or CAD file types that Fusion supports. As Autodesk regularly updates Fusion, you never have to worry about not having access to the latest CAD translators:

All supported files (*.wire;*.dwg;*.f3d;*.f3z;*.fsch;*.fbrd;*.flbr;*.iam;*.ipt;*.(

All supported files (*.wire;*.dwg;*.f3d;*.f3z;*.fsch;*.fbrd;*.flbr;*.iam;*.ipt;*.
Alias files (*.wire)
AutoCAD DWG file (*.dwg)
Autodesk Fusion 360 files (*.f3d;*.f3z)
Autodesk Inventor files (*.iam;*.ipt)
Catia V5 files (*.CATProduct;*.CATPart)
DXF files (*.dxf)
FBX files (*.fbx)
IGES files (*.iges;*.ige;*.igs)
NX files (*.prt)
OBJ files (*.obj)
Parasolid Binary files (*.x_b)
Parasolid Text files (*.x_t)
Pro/ENGINEER and Creo Parametric files (*.asm;*.prt)
Pro/ENGINEER Granite files (*.g)
Pro/ENGINEER Neutral files (*.neu)
Rhino files (*.3dm)
SAT/SMT files (*.sab;*.sat;*.smb;*.smt)
SolidWorks files (*.prt;*.asm;*.sldprt;*.sldasm)
STEP files (*.ste;*.step;*.stp)
STL files (*.stl)
SketchUp files (*.skp)
Solid Edge files (*.par)
T-Spline files (*.tsm;*.tss)
123D files (*.123dx)
3MF files (*.3mf)

Figure 1.5 – A list of the available CAD and MESH file types that Fusion can open

This is a great option to open a single file quickly, but there are several downsides to this option:

- You can only open one file at a time using this method.

- When using this method, you end up in **Direct Modeling mode** upon opening a CAD file. If you want to capture design history, you will need to manually turn on the design history using the browser. Modeling with the **Design history** and **Direct modeling** methods will be covered in detail in *Chapter 2*.

- You do not get a chance to set the units or part placement when opening STL/OBJ files.

In conclusion, if you try to open STL or OBJ files, this is not the option I would recommend. However, if you open a CAD file (for example, STEP or SAT) or a 3MF file, this is a quick and easy option that you may want to use. Don't forget to manually turn on **Capture Design History** if you want to store your edits in the timeline, as shown in *Figure 1.6*:

Figure 1.6 - After a File | Open... workflow, turn on Capture Design History via the browser

Another option available when opening CAD and Mesh files is within Fusion's **Data Panel**. You can access the **Data Panel** window by selecting the **Show Data Panel** icon, located in the top-left corner of the interface, as shown in *Figure 1.7*.

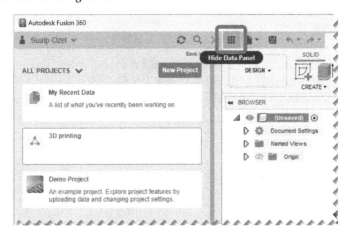

Figure 1.7 - Show/Hide Data Panel

Once the data panel is visible, you can navigate to the team and project where you wish to upload your design data. Once in the appropriate project, you can create new folders to organize your designs and documents further. Once you are in the desired project/folder location, you can use the **Upload** button, as shown in *Figure 1.8*, and select or drag and drop files to the upload dialog that is on your screen. Either option will allow you to upload multiple CAD or mesh files to the designated team/project/folder location. Please note that each uploaded file will become its own Fusion design document.

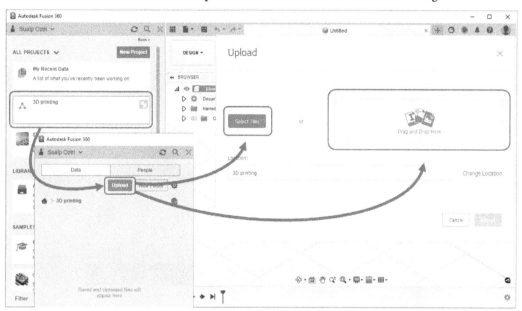

Figure 1.8 – Uploading multiple Mesh and CAD documents to Fusion using the Data Panel

This is a great option to upload multiple CAD models to Fusion. However, just like the previous option, you will have to turn on **Capture Design History** if you want to store your edits in the timeline, as shown in *Figure 1.6*. And just like the previous option, this method is not one I would recommend opening STL or OBJ files, as it does not give you initial control over units of part placement during the import process.

Another lesser known but rather useful option is based on the Fusion mobile app. If you have a mobile device, you can download the Fusion app from the relevant app store, and you can browse for and upload CAD or Mesh files stored on your phone or tablet, as shown in *Figure 1.9*:

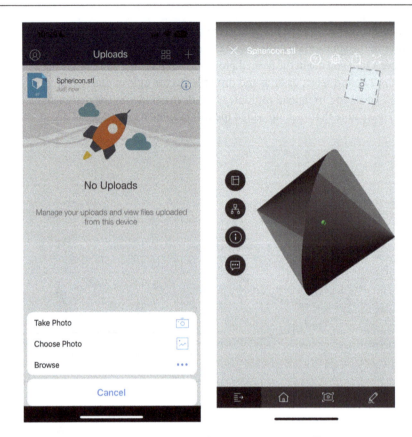

Figure 1.9 – A view on a mobile device after uploading an STL file to Fusion Team

This method is particularly useful if you do not have your computer available when you receive a CAD or mesh file. In such instances, you can quickly upload the design to your Fusion Team/project/folder so that you and your team members can start working on the design right away. Just like the previous methods, this method also requires that you manually turn on the **Capture Design History** within Fusion's browser if you want to store your edits in the timeline, as shown in *Figure 1.6*.

> **Important note**
>
> If you upload an STL/OBJ file to Fusion Team or the Fusion app, after opening it in Fusion, you will have to inspect the units and the position of the mesh body to make sure it matches the original design intent. Fusion uses centimeters as the default unit system when creating Fusion design documents from mesh file types with no units.
>
> For example, imagine an STL file that represents a sphere with a diameter of 10 mm. As STL files do not contain unit data, if we use the upload workflow on this STL, as shown in *Figure 1.8*, to Fusion Team, the resulting Fusion design document will be a sphere with a diameter of 10 cm.

Now that we have learned about the various ways we can open and upload both CAD and mesh files, it's time to start focusing on workflows specific to mesh bodies. In the next section, we will learn how to insert mesh files directly into our Fusion design documents.

Inserting Mesh workflows for STL, OBJ, and 3MF files

Even though we can upload mesh files to Fusion Team and simply open them in Fusion after the upload is complete, my personal preference when opening mesh files is the **Insert Mesh** command located within the **DESIGN** workspace | the **Mesh** tab | the **CREATE** panel. Using the **Insert Mesh** command, you can insert one or multiple mesh bodies into an active Fusion design. You do not need to save the design as a Fusion document before inserting the mesh files. This is an important distinction, as a similar workflow to insert all Fusion bodies/components, which rely on the **Insert Derive** function (as shown in *Figure 1.10*) located within **DESIGN** | **Solid**, **Surface**, **Mesh**, **Sheet Metal**, and **Plastic** | **Insert Derive**, require a Fusion design to be saved before inserting externally referenced Fusion designs into an active design:

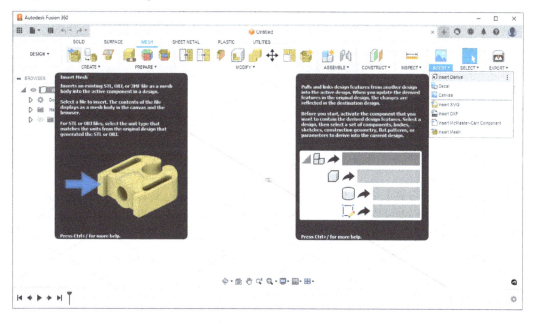

Figure 1.10 – Insert Mesh versus Insert Derive

If you are inserting a single mesh (STL or OBJ) file, you will have the ability to edit the location and units for that mesh file at the time of insertion. If you import multiple mesh (STL or OBJ) files, the location and unit data you edit during the insert operation apply to all the files you insert simultaneously. Because of this, my preferred method is to insert the mesh files individually so that I can visually inspect them and edit their location and units one at a time, especially if I insert mesh files created by different designers, as they may have used different unit systems and coordinate systems.

Now, let's focus on the **Insert Mesh** command and some of its options. For this first exercise, we will insert the CLIP.STL file. After initiating the **Insert Mesh** command and selecting the STL file, you will see the **INSERT MESH** dialog, with options to change the units and reposition the part. The **Flip Up Direction** button rotates the part around the *X* axis by 90 degrees. The **Center** button moves the inserted mesh by referencing the center of its bounding box to the origin of the active component. The **Move To Ground** button translates the mesh body so that its lowest point touches the ground plane. In *Figure 1.11*, the ground plane is the *XY* plane:

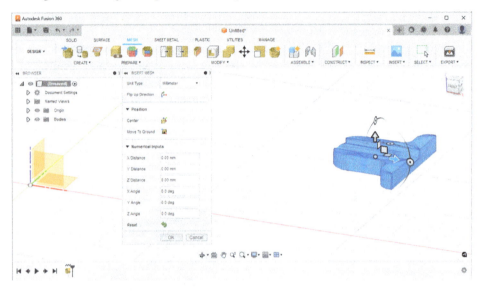

Figure 1.11 – The INSERT MESH dialog

In addition to centering and grounding a mesh body, we can also use the in-canvas manipulators (arrows and rotation tools) or the numerical inputs section within the **INSERT MESH** dialog to reposition the mesh body until we are satisfied. We can always revert our actions with the **Reset** button.

In this example, let's first use the **Center** command, then the **Move To Ground** command, and press **OK**.

Important note

The default model orientation can be set in Fusion's **Preferences** dialog in the **General** section. This book will rely on the *Z up* orientation for all examples.

The default units for every new design can be set in Fusion's **Preferences** dialog in the **Default Units** section's **Design** subsection. This book will rely on millimeters as default units for all new designs.

The default state of design history can be set in Fusion's **Preferences** dialog in the **General** section's **Design** subsection. This book will rely on **Capture Design History** (parametric modeling) for all new designs.

All these preferences can be seen in *Figure 1.12*:

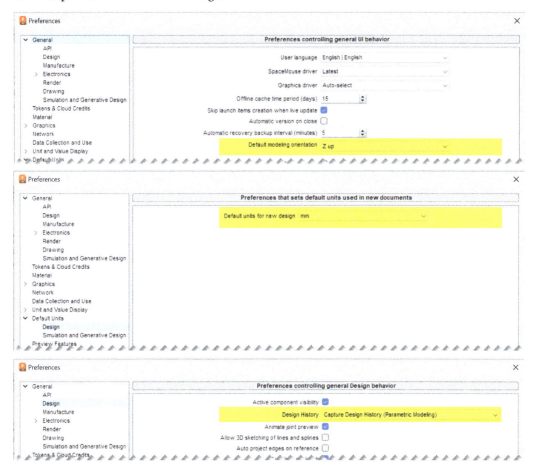

Figure 1.12 – Various sections within Fusion's Preferences dialog

After we finish inserting this STL file, we will be left with a **base mesh feature** added to the timeline, and the mesh body located on the *XY* plane will be centered around the origin, as you can see in *Figure 1.13*:

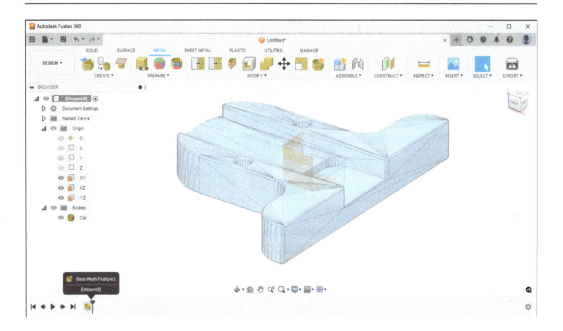

Figure 1.13 – The Insert Mesh command completed

Unlike STL files, 3MF and OBJ files can also contain color information. Now, let's insert the CLIP.3MF file using the same **Insert Mesh** command into a new design file. In Fusion, we can create a new design file by selecting the + icon to the right of our list of open Fusion design documents. After inserting a 3MF file, Fusion shows the mesh body with a unique color for each face group. A face group is a collection of faces that typically represent a feature or a region on a mesh body. This 3MF file is made of a single face group. Therefore, we will see a single color on all the faces after inserting this file into Fusion. We can visualize the color information contained within the 3MF and OBJ files by toggling face group visualization, using the **Display Mesh Face Groups** command in the **INSPECT** panel, as shown in *Figure 1.14*:

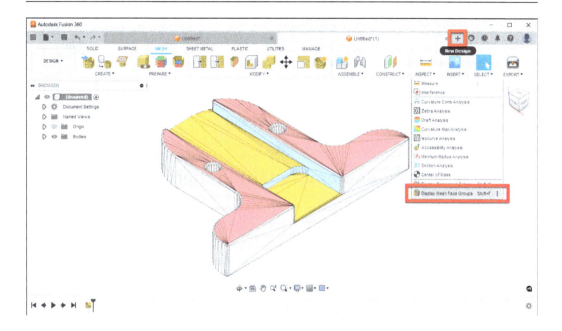

Figure 1.14 – 3MF and OBJ files can contain color

Important note

As shown in *Figure 1.15*, the **Apply a different appearance** checkbox must be unchecked in Fusion's **Preferences** dialog | **General** section | **Material** subsection, in order to display color information contained within a 3MF or OBJ file opened in Fusion, using the **Insert Mesh** workflow previously described.

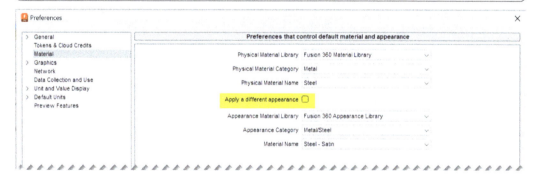

Figure 1.15 – The Material subsection in Fusion's Preferences dialog

In this section, we learned about the various methods we can use to insert mesh files such as STL, OBJ, and 3MF. We also covered how to set units and relocate mesh bodies during the insert workflow. In addition, we went over how to visualize the color/texture data for those files.

Now, it is time to start focusing on workflows specific to inspecting mesh bodies we insert for potential problems and repairing them. In the next section, we will learn how to utilize the automatic inspection and repair functionality within the **Mesh** tab and start exploring how to manually repair Mesh bodies when the automatic repair is not enough.

Inspecting Mesh bodies and repairing them

STL is a simple file format and is widely used in 3D printing. Almost all 2D and 3D design software has been able to create STL files since its invention back in 1987. Because of its simplicity, long history, and wide availability, most STL files in existence contain errors and are generally in need of repair before they can be used for manufacturing.

In this section, we will learn how to inspect the mesh bodies we insert into Fusion and how to repair broken STL files using automatic and manual repair techniques, in order to prepare them for 3D printing.

To demonstrate the various methods we can use to inspect and repair mesh bodies, we will use an STL file called `Clip_Broken.STL`. Let's start by creating a new Fusion design, as shown in *Figure 1.14*.

Now, we can use the **Insert Mesh** command and insert the `Clip_broken.STL` file into our active Fusion design document. Just like the last example, let's utilize the **Center** and **Move To Ground** functionalities in the **Insert Mesh** dialog consecutively to translate the part closer to the origin and above the *XY* plane. Now, we can look at the details of the mesh body we just inserted.

Fusion automatically inspects all mesh bodies inserted and displays a warning icon next to any problematic mesh body. As shown in *Figure 1.16*, if you expand the bodies list in the browser, you will see a mesh body named `Clip_Broken`, and next to it, you will see a warning icon. If you hover over the warning icon, you will see the following information about the mesh body:

- **Mesh is not closed.**
- **Mesh is not oriented.**
- **Mesh does not have positive volume.**

When we see such information, we often need to take repair actions before we can proceed to 3D-print the files:

Figure 1.16 – A warning icon next to a problematic mesh body

Another useful tool for inspecting problematic mesh bodies is the **PROPERTIES** dialog. If you select a given mesh body, right-click, and select **Properties**, you can see a detailed list of information about the mesh body, as shown in *Figure 1.17*.

In this example, you can see that this mesh body does not have a mass or volume, even though it has been automatically assigned a physical material (**Steel**) with a density. This is another indication that this mesh body does not enclose a volume. Refer back to *Figure 1.15*, which shows the logic behind the automatic material assignment within the **Material** subsection of the Fusion **Preferences** dialog.

> **Important note**
>
> A common term used in 3D printing to describe bodies that properly enclose a volume is *watertight*. Having a watertight body is a desirable feature when attempting to 3D-print objects. However, just because an object is not watertight does not mean it is not 3D-printable. In metal powder bed fusion 3D printing, support structures are often made up of surfaces (not watertight bodies). We will cover this topic in more detail in *Chapter 12*.

If you expand the **Mesh** subheading, you will see that this mesh body has two shells. A shell is a collection of triangles (and sometimes quadrilaterals) that are connected to one another. If a mesh body is made up of multiple shells, it often indicates a broken continuity of the triangles that make up a mesh body.

Sometimes, a mesh body can have multiple shells and still be valid. For example, imagine a hollow sphere. In such a case, you would have one shell that represents the outside of the sphere and another shell that represents the inner boundary.

If you expand the **Mesh Analysis** subsection, you will see that this mesh body is not closed, not oriented, and does not have a positive volume:

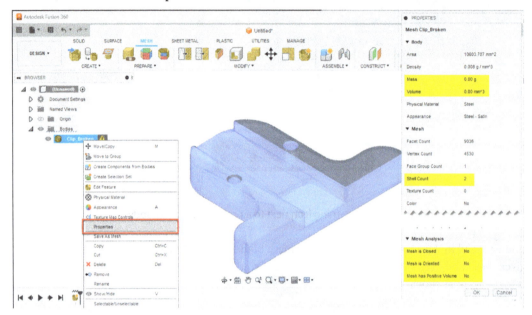

Figure 1.17 – The PROPERTIES dialog shows the relevant mesh analysis information

Now that we understand we have a problematic mesh body on our hands, we can close the **PROPERTIES** dialog, and select the **Repair** icon on the ribbon in the **MESH** tab | the **PREPARE** panel.

Fusion has four automatic repair types. Let's start by selecting the **Close Holes** repair type, which fixes flipped triangles and closes holes. If we activate the preview checkbox within the **REPAIR** dialog and expand the detailed analysis, we can edit the sliders to see the impact of the selected repair type on the body we wish to repair. *Figure 1.18* shows that this repair type will automatically fix the part so that it can be 3D-printed, but a visual inspection of the outcome looks cluttered, with many triangles overlapping each other.

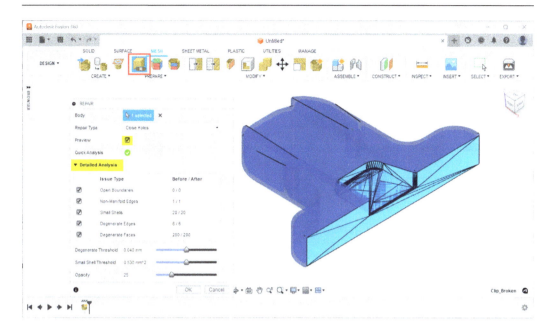

Figure 1.18 – The Close Holes repair type and its impact on this part

Even though the outcome of this automatic repair is reported as a success, it is always a good idea to perform a visual check and review the details of the mesh body properties. *Figure 1.19* shows the **PROPERTIES** dialog for the **Clip_Broken** mesh body after we performed the **Close Holes** repair type. Here, we can see that this mesh is made up of 66 shells. This is an unexpected outcome, and a visual inspection of the part shows problematic zones, as highlighted in red ovals. This part is a symmetrical part along the *XZ* plane, yet the automatic repair closed one of the holes instead of creating the desired countersink hole type outcome:

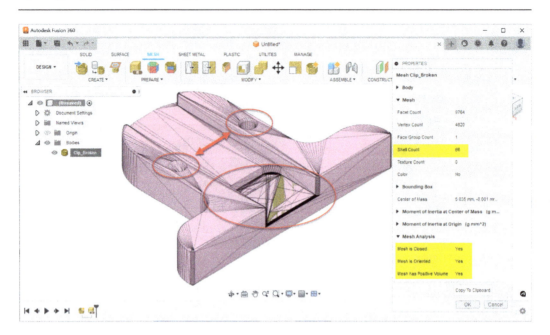

Figure 1.19 – The PROPERTIES dialog shows the outcome of repaired mesh body

Now, let's use the **Edit Feature** command on the repair action we just performed, by selecting it from the timeline at the bottom of Fusion's user interface and right-clicking on it, as shown in *Figure 1.20*. This time, we can change our repair type to **Stitch and Remove**. This repair action includes all the repairs that the previous repair type executed, and stitch triangles remove double triangles and degenerate faces, as well as tiny shells. Using this repair type takes a little bit longer, but the outcomes look much better. As you can see in *Figure 1.20*, we no longer look at a cluttered mesh in the region highlighted with the large red oval. However, we still have the same issue of not having the desired outcome of the countersink hole:

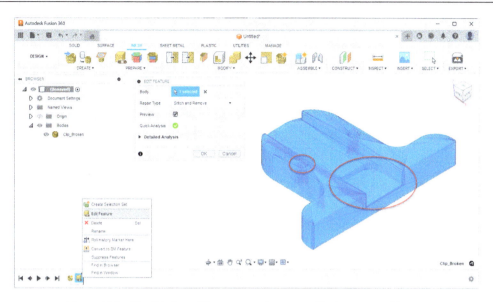

Figure 1.20 – The Stitch and Remove repair type and its impact on this part

Before we move on to the next repair type, let's take a closer look at our part by adding a **section analysis**. To add a section analysis, we must use the **Section Analysis** command located in the **INSPECT** panel. After we select the cut plane as the *XZ* plane and flip the segment we want to section, we can easily see that this mesh represents a hollowed-out object, as shown in *Figure 1.21*. Looking at our part with a section analysis, it makes more sense why we saw a shell count of two within the **PROPERTIES** dialog:

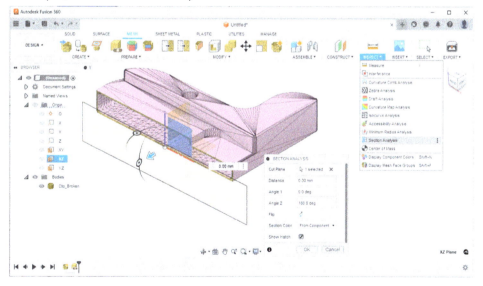

Figure 1.21 – A section analysis shows this mesh body is a hollow object

While the section analysis is visible, as denoted by the gray eye icon next to **Section1** in the browser, we can edit our repair from the timeline and utilize the third repair type, called **Wrap**. This repair type makes the same repairs as **Stitch and Remove** and wraps the surface of the mesh body, removing all inner structures. As you can see in *Figure 1.22*, the outcome of this repair type removes the secondary shell inside our part, making our body a solid object. In this example, our goal was not to make such a drastic change to this part. Therefore, we will edit the repair one more time and move on to the *fourth* option. However, in some cases where the inner shell is also in need of major repair, this repair type could easily remove the triangles that belong to the inner shell, speeding up the mesh repair steps:

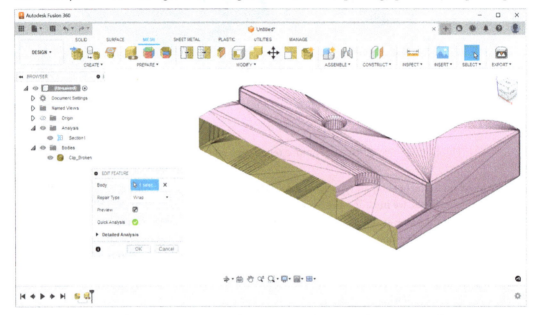

Figure 1.22 – The Wrap repair type and its impact on this part with a section analysis view

The fourth and final repair type is called **Rebuild**. There are multiple rebuild types available, and in *Figure 1.23*, you can see the outcome of a fast rebuild type with an average mesh density. This repair type reconstructs the mesh body and is the slowest repair type based on your density input. It can be used as a last resort when automatically repairing parts. However, the mesh outcome of this repair type has a high triangle count, and this repair type does a less desirable job of capturing details around sharp edges.

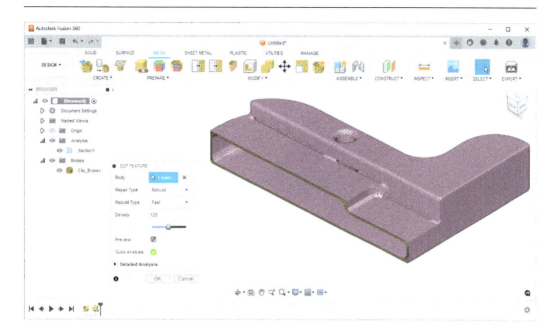

Figure 1.23 – The Rebuild repair type and its impact on this part with a section analysis view

Now that we have covered all four repair types, let's edit a repair action, reuse the **Stitch and Remove** repair type, as its outcome was the most desirable for this part. This repair type also happens to be the default repair type recommended by Fusion.

None of the repair types were able to detect and repair the missing surface around the countersink hole. It looks like we will have to do some manual mesh editing actions to fix this issue. Let's start by cutting the part in two, using the **Plane Cut** feature located in the **MODIFY** panel. The **Clip_Broken** mesh body is the body we will select to cut, and the cut plane will be the *XZ* plane. We will use the **Trim** type to remove the unwanted section and use the **Minimal** fill type to turn the remaining mesh body into a watertight object.

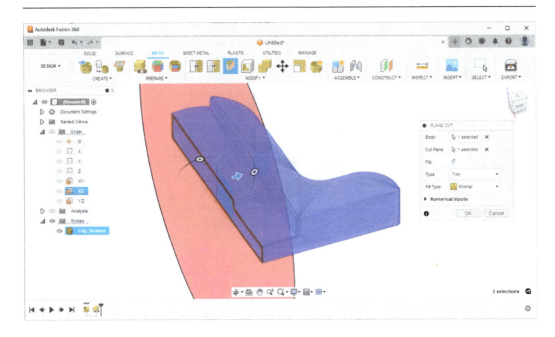

Figure 1.24 – Trimming the mesh body using the Plane Cut tool

Once we have cut the mesh body in half, we will have to create a component from the mesh body in order to mirror it. *Figure 1.25* shows how to create a component from a trimmed mesh body – by right-clicking on it in the browser and selecting the **Create Components from Bodies** option.

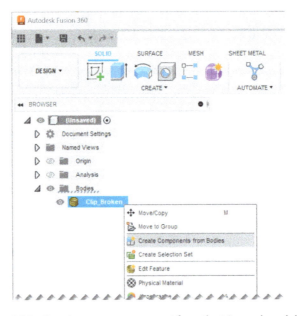

Figure 1.25 – Creating a new component from the trimmed mesh body

Once we have a component, we can toggle to the **SOLID** tab. We can select the **Mirror** command located under the **CREATE** panel. Once the **MIRROR** dialog is visible, we will have to change our object type to **Components** and select the newly created component to mirror. Our mirror plane will once again be the *XZ* plane. The outcome of the mirror action can be seen in *Figure 1.26*:

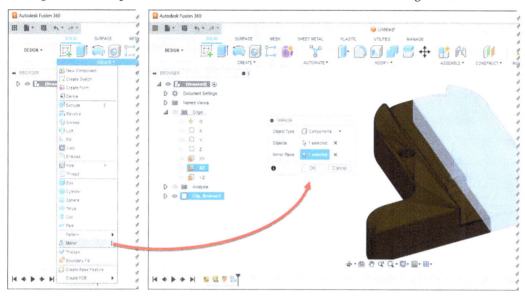

Figure 1.26 – Mirroring a component

After we mirror our component, we will end up with two mesh bodies within two components. It is time to combine them into a single unified component. To accomplish this, let's go back to the **MESH** tab and utilize the combine command located within the **MODIFY** panel. We will have to select a target body and a tool body, which will be the **Clip_Broken** mesh bodies from the two components. We will utilize the join operation, create a new component, and keep the tools. This will ensure that we can easily compare the outcome and the input if desired in the future. *Figure 1.27* shows the necessary input to combine two mesh bodies to make a new component:

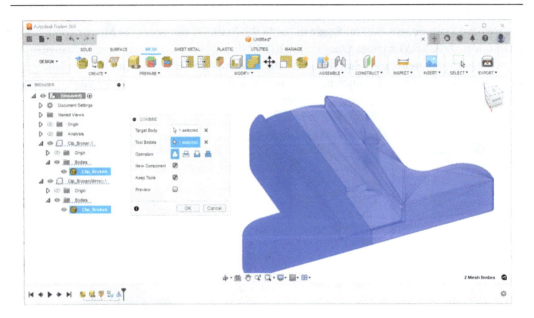

Figure 1.27 – Combining mesh bodies to make a new component

The outcome of the combined operation is a fully repaired body in a new component named Component3 in the browser. *Figure 1.28* shows this new component while hiding the section analysis, as well as the first and second components:

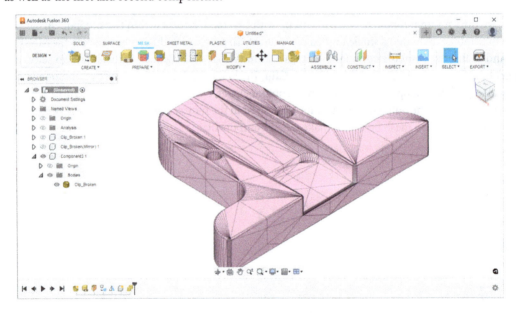

Figure 1.28 – Component3, a fully repaired mesh

If we turn on the visibility for the section analysis, we can quickly see that the mesh body we repaired still has a hollow cross-section and can be 3D-printed, as shown in *Figure 1.29*:

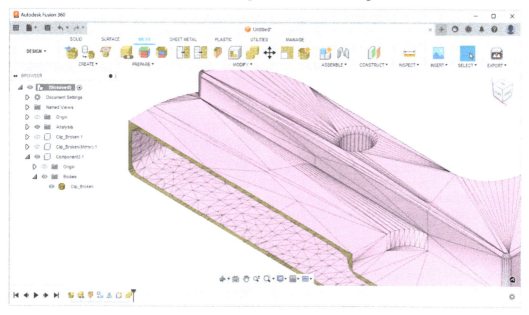

Figure 1.29 – A section analysis view of the repaired mesh body

As a final sanity check, it is always a good idea to look at the properties of the repaired mesh body to make sure we have a closed and oriented mesh with a positive volume. The timeline of all our actions is captured in the design history in the bottom-left corner of the browser, as shown in *Figure 1.30*. You can always replay those steps in the future and remind yourself how you were able to repair this mesh.

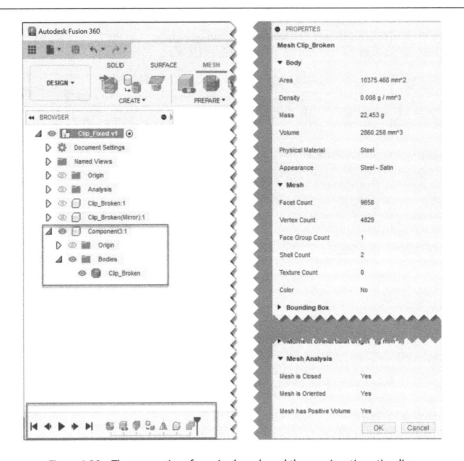

Figure 1.30 – The properties of repaired mesh and the repair actions timeline

The Git repository for this chapter also includes a Fusion file called `Clip_fixed.f3d`, which you can open directly in Fusion to replay the timeline if you had any issues during this example.

Summary

In this chapter, we learned about the various ways to open CAD and mesh files in Fusion, ranging from utilizing the **File | Open...** workflow to uploading files to Fusion Team from a web browser or the Fusion desktop and mobile apps.

We also covered how to insert mesh files into our active designs, how to set the units, and how to position those parts. We highlighted how to render and color data contained in certain mesh files. We talked about how to inspect problematic mesh bodies. We also went through an example using both automatic and manual repair steps, such as trimming mesh bodies and mirroring components, to fully repair a heavily damaged mesh file.

Using the skills you learned in this chapter, you can now feel confident in your ability to open, inspect, and repair most mesh files using Fusion.

In the next chapter, we will go over how and why we may want to work with parametric modeling versus direct modeling. We will also cover how to automatically detect features for models we import to Fusion and how we can edit those features with Design for Additive Manufacturing principles.

2

Editing CAD/Mesh Files with DFAM Principles in Mind

Welcome to *Chapter 2*. In this chapter, we will cover how to edit CAD/Mesh models with various **Design for Additive Manufacturing** (**DFAM**) principles in mind. We will start by highlighting how to switch on/off the design history. We will demonstrate the differences in workflow when using parametric versus direct modeling. We will then demonstrate how to automatically detect features for imported CAD models. Then, we will go over how to edit both CAD and mesh bodies based on DFAM principles. We will conclude the chapter by showcasing how to convert mesh files into solid bodies that can be further modified for manufacturing.

In this chapter, we will cover the following topics:

- Introduction to design history with Fusion 360

- Recognizing CAD features and editing them

- Working with Mesh files natively in Fusion 360

By the end of this chapter, you will have learned how to work with models using parametric and direct modeling workflows in Fusion 360. You will be able to automatically recognize CAD features and edit them. You will also know how to edit CAD/mesh geometry imported from other sources using Design for Additive Manufacturing principles.

Technical requirements

Most of the topics covered in this chapter are applicable to all licensing levels (personal/hobby, startup, educational, and commercial) of Fusion 360. However, the book will also cover advanced functionality not available to personal/hobby use licenses. Such advanced functionality will be identified as *N/A for Personal Use* and *Requires Access to Extension* as appropriate.

The lesson files for this chapter can be found here: `https://github.com/PacktPublishing/3D-Printing-with-Fusion-360`.

Introduction to design history with Fusion 360

In *Chapter 1*, we briefly touched on how Fusion 360 can capture the design history of model changes. In this chapter, we will dig a little deeper into this topic and showcase some of the benefits and drawbacks of capturing design history (parametric modeling) versus not capturing design history (direct modeling).

In the context of Fusion 360, Parametric Modeling is a methodology based on features and constraints. All features you add to a model are captured over time and any changes you make to a given feature propagate downstream in your timeline and impact the subsequent features and ultimately the final outcome of your design. This makes parametric modeling a powerful tool for creating design alternatives based on rules.

In contrast, direct modeling is a methodology where the features you add to your model are not all stored or editable. A subset of direct modeling features can be edited. However, those features do not interact with any feature you may have added prior to or after the one you are editing. This makes direct modeling a very flexible tool for quickly making drastic changes to a design. With direct modeling, you don't have to worry about any issues a given change might cause, based on how your model was originally designed in the CAD software.

When generating a new design, Fusion 360 relies on the setting designated in the **Design History** option within the **Preferences** dialog, as shown in *Figure 2.1*, to determine whether the design will utilize Parametric or Direct modeling methods.

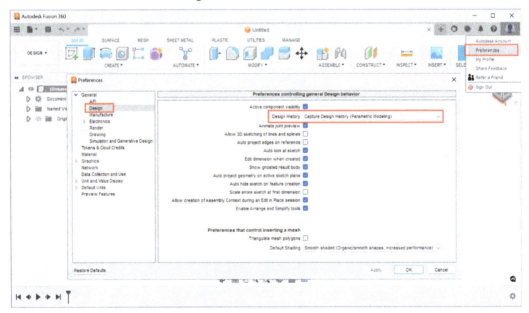

Figure 2.1 – Fusion 360's preferences for capturing design history

When working on an existing Fusion 360 Design, we also have the option to change our minds and utilize direct modeling instead of parametric modeling at any point in time. We have two ways we can use to change from Parametric modeling to Direct modeling. The first way to switch from parametric modeling to direct modeling is by right-clicking the first item on the browser and selecting **Do not capture design history**. The second way to switch is by selecting the gear icon in the right corner of the Fusion 360 timeline and selecting **Do not capture design history** from the list of available options. Both methods are shown in *Figure 2.2*.

> **Important note**
>
> If you switch from parametric modeling to direct modeling by mistake, you can always undo your actions (*Ctrl + Z*) and get back your timeline.

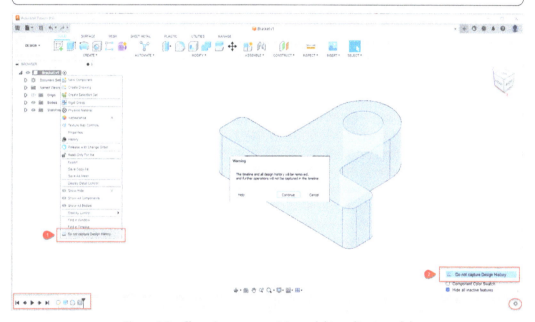

Figure 2.2 – Changing a parametric model to a direct model

Regardless of the method you use, when switching from parametric modeling to direct modeling, the timeline of your features will be lost, the timeline bar will be invisible, and you will lose the constraints and references between the features you used to create your design. *Figure 2.3* shows the impact of switching from parametric modeling to direct modeling. Notice how the figure on the left has a timeline bar at the bottom. The timeline bar shows how the parametric modeling methodology was used to create this part. All the listed features have the dimensions and references used when generating them.

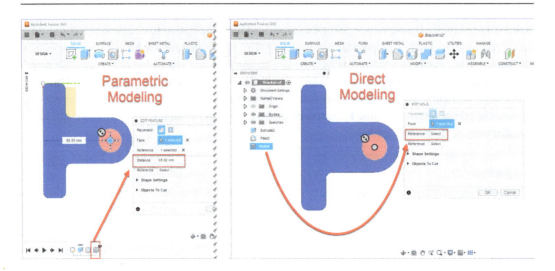

Figure 2.3 – Impact of switching from parametric modeling to direct modeling

After switching to the direct modeling methodology, the timeline bar is lost. The model now has a list of selectable features; some of them (for example, **Fillet** and **Hole**) are even editable. However, if you edit such a feature, you will see that the references will not be saved with the feature. With every subsequent edit, you will be able to redefine those references. In other words, features are not constrained and can be freely moved around, even after being defined explicitly.

Parametric Modeling

Fusion 360 has several pieces of functionality that are only accessible when you are in parametric modeling mode. The most obvious one is the fact that you can define, use, and edit parameters that control inputs for features. *Figure 2.4* shows how a list of named and user-defined parameters were utilized in the creation of this model. If you are making drastic changes to your component and/or assemblies, you may need to utilize the **Compute All** command to resolve any issues in the timeline. Just like the **Change parameters** command, the **Compute All** command is only available when you are working within the parametric modeling paradigm. Both commands can be found in the **MODIFY** panel, as shown in *Figure 2.4*.

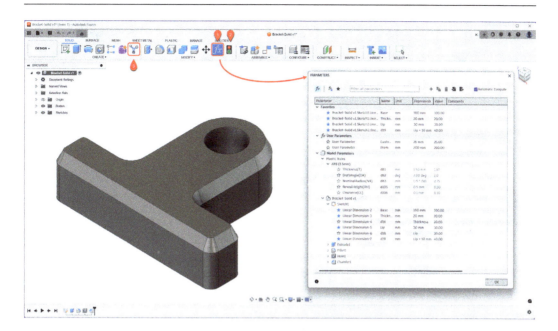

Figure 2.4 – Change Parameters, Compute All, and Automated Modeling commands

Another critical piece of functionality that is unique to Fusion 360's Parametric modeling toolset is **Automated Modeling**, located in the **AUTOMATE** tab, as shown in *Figure 2.4*. **Automated Modeling** connects disjointed faces within a Fusion 360 design and creates multiple alternative bodies/components you can choose from. We will cover **Automated Modeling** in detail in *Chapter 3*.

There are three simplification features (**Simplify**, **Remove Faces**, and **Replace with Primitive**) that you can only utilize when you are in Parametric modeling mode. But you have alternative commands you can use when in direct modeling mode to achieve similar outcomes. In Parametric modeling, you can access these three from the **MODIFY** panel | the **Simplify** slide-out menu, as shown in *Figure 2.5*.

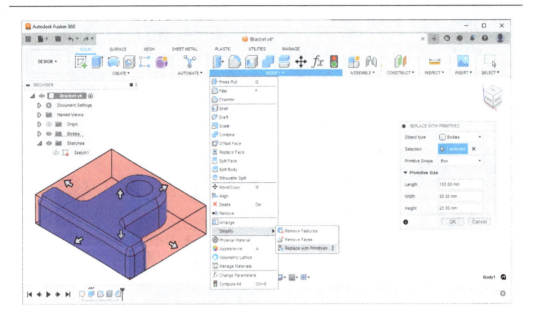

Figure 2.5 – Simplify functions available with the Parametric modeling method

Unlike most other parametric modeling functions, these three are not editable once used. If you use them, you will have to roll back the timeline or delete the feature from the timeline in order to undo their effect, as shown in *Figure 2.6*:

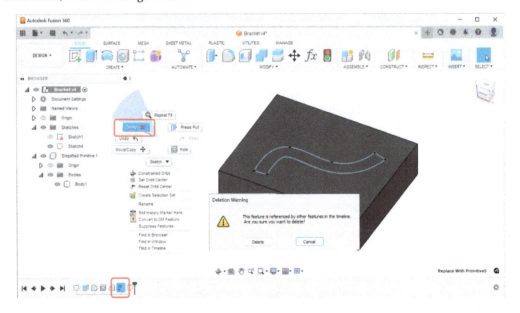

Figure 2.6 – Deleting a Replace with Primitives feature referenced by a Sketch

Whenever you are in parametric modeling mode and you delete features (or bodies or components) in the middle of a timeline, you run the risk of breaking the timeline continuity. If you delete an item from the timeline that is used by other items as a reference, you will be left with missing reference warnings. In the example shown in *Figure 2.7*, once the **Replace with Primitives** feature is deleted from the timeline, we are left with a yellow highlighted timeline icon for the **Sketch** feature, which used the top face of the now-deleted simplified primitive body.

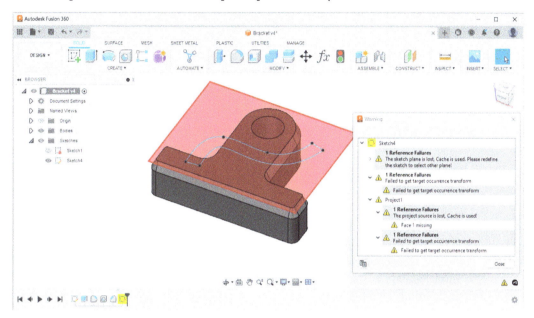

Figure 2.7 – Reference failure warnings in the timeline after deleting a referenced feature

Having a component with reference failure warning messages is generally not a desirable outcome as we may not be able to replay the timeline or compute all changes effectively. Features with such warnings will require additional work to reestablish lost references. Therefore, you may choose to avoid using these three features altogether. As shown in *Figure 2.8*, you can disable **Arrange and Simplify tools** within Fusion 360's **Preferences**, by navigating to the **General** pull-down menu | the **Design** subsection.

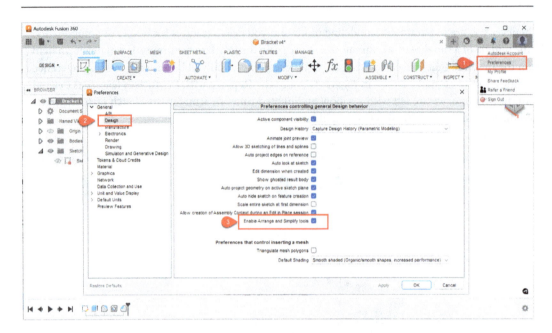

Figure 2.8 – Enable/disable simplify tools within Fusion 360's Preferences

As we just illustrated, using the **Delete** command can create unforeseen consequences when we are in parametric modeling mode. There are two alternatives to the **Delete** command you can use to eliminate unwanted features of bodies. They are the **Suppress Features** command and the **Remove** command. Both are unique to Parametric modeling and are my personal preferred way to exclude features and models from Fusion 360 design documents without breaking the timeline.

As shown in *Figure 2.9*, the **Remove** command eliminates bodies and components and can be accessed by selecting its input and using the right-click menu. *Figure 2.9* also shows the **Suppress Features** command, which allows us to exclude features from the model and can be accessed by selecting its input from the timeline and using the right-click menu.

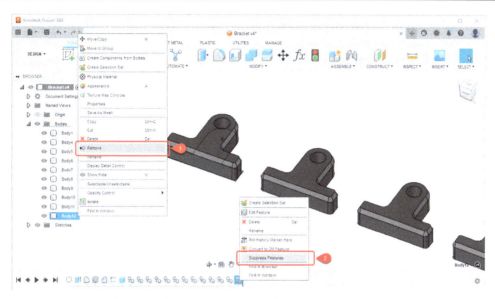

Figure 2.9 – The Remove command and Suppress Features

The final unique set of features for parametric modeling *requires access to the Design extension.* Fusion 360 offers several extensions to trial users, educational users, startup users, and commercial users. Extensions are not available to free, personal, or hobbyist users. Depending on your subscription type, you can gain access to extensions within the **Extensions** dialog, as shown in *Figure 2.10.*

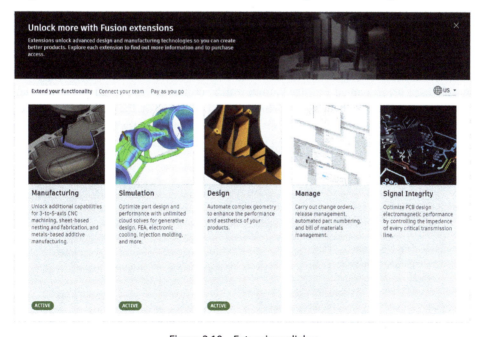

Figure 2.10 – Extensions dialog

Once you gain access to the **Design** extension, you can assign and edit plastic rules to components within the plastic tab's setup panel. As shown in *Figure 2.11*, assigning a plastic rule to a given component automatically adds a handful of parameters to Fusion 360's Parameter table, which controls settings such as **Thickness, Draft Angle, Nominal Radius**, and **Knife Edge Threshold**.

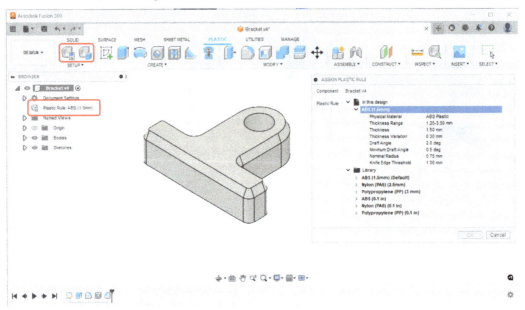

Figure 2.11 – Assigning a plastic rule to a component

After assigning a plastic rule to a given component, any plastic feature that you add to that component automatically utilizes parameters set by the plastic rule. In *Figure 2.12*, we can see a shelled version of our design with added webs. The **Web** feature is not unique to parametric modeling. But it is currently utilizing a preset named **Feature Values**, which automatically assigns a parameter named **T** for **thickness** and **DA** for **draft angle**. It also uses an equation based on the thickness parameter for the fillet radius. All these parameters and the feature preset are based on the plastic rule we assigned in the previous step. Therefore, these modeling techniques are all dependent on using the parametric modeling methodology.

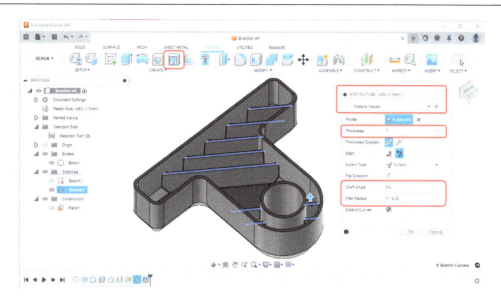

Figure 2.12 – Creating/editing a Web feature using a plastic rule

The final feature I want to highlight, which is unique to the parametric modeling methodology, is **Geometric Pattern**. You can access **Geometric Pattern** from the **CREATE | Pattern** slide-out menu. *Figure 2.13* shows the outcome of a geometric pattern applied to a face on our model. Using this feature, you can parametrically create a modeled surface texture with ease. Just like the plastic rule and feature presets, **Geometric Pattern** also requires access to the Design extension.

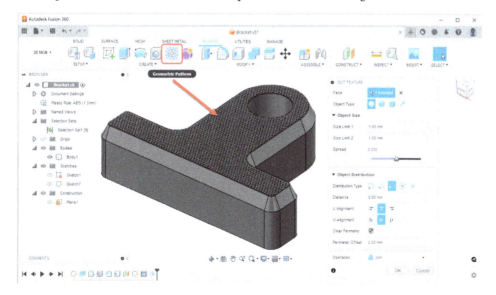

Figure 2.13 – Creating/editing a geometric pattern

Direct modeling

Fusion 360 has several pieces of functionality that are only accessible when you are in Direct modeling mode. Some of this functionality is specific to working with mesh bodies. Some of it is related to working with Solid and Surface bodies and some of it is related to working with components within an assembly.

Let's start by covering the Mesh-specific functionality. The first mesh functionality specific to direct modeling is **CREATE FACE GROUP**, located under the **MESH** tab | the **PREPARE** panel. Using **CREATE FACE GROUP**, you can select individual triangles or a group of them and assign them to a new face group. The creation of face groups is crucial in order to select them downstream during additive manufacturing workflows for support structure creation or when attempting to convert mesh bodies into solid objects. Even though Fusion 360 has tools for generating face groups automatically, having manual control over creating face groups is an important piece of functionality when dealing with complex inserted meshes. You can use this functionality as a part of direct modeling.

> **Important note**
>
> In parametric modeling, you will need to utilize the **DIRECT EDIT** feature to be able to create face groups manually. All the mesh-specific functionality we are highlighting in this subsection can be accessed by introducing the same direct edit feature within parametric modeling.

The next mesh-specific functionality for direct modeling is **ERASE AND FILL**. This command is located under the **MODIFY** panel and allows you to remove and replace one or more faces on your mesh body to manually fill holes and repair regions. When using this command, I also recommend turning on **MESH SELECTION PALETTE** so you can utilize different selection types, such as the paintbrush or free-form selection. If you're using the paint selection, you can also control the brush size. The **Mesh Palette** is accessible from the **SELECT** panel, as shown in *Figure 2.14*.

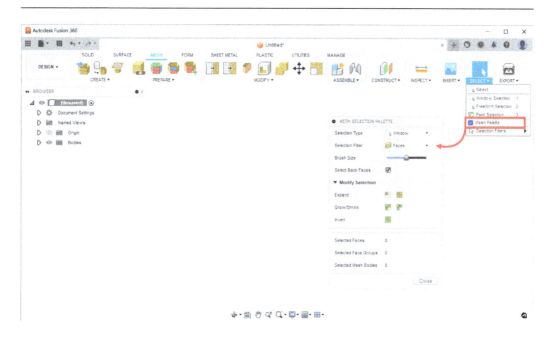

Figure 2.14 – Mesh selection palette and mesh selection tools

Now, let's start focusing on solid and surface body-specific features that are unique to direct modeling. The first feature we will highlight is generally applicable to solid and surface bodies converted from mesh bodies using the faceted method. We will cover how to convert mesh bodies into solid bodies later in this chapter when we talk about working with Mesh files natively in Fusion 360. Once a mesh body is converted into a solid body using the faceted method, it will have a large number of surfaces, which will make the design difficult to work with due to its file size and model complexity. It is generally a good idea to simplify such models in order to further edit them or use them in an assembly. In such cases, we can rely on the **MERGE** feature, located within the **SURFACE** tab | the **MODIFY** dropdown. This feature will allow you to combine two or more faces within a solid or surface body into a single face. If the **Select Chain** checkbox is activated, you can simply double-click on a single face and Fusion 360 will automatically select all faces that are connected to the face you have selected that are coplanar, as shown in *Figure 2.15*.

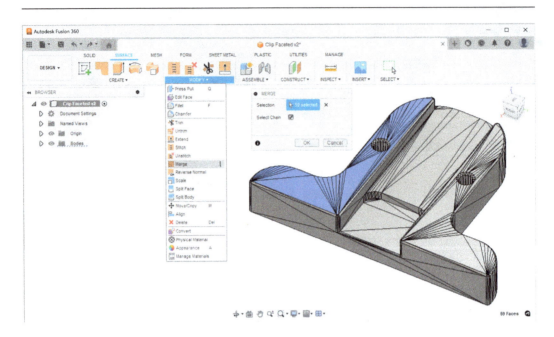

Figure 2.15 – Merge planar faces to create a single face

After pressing **OK** on the **MERGE** dialog, all the selected faceted faces will be replaced with a single face, reducing the model complexity and file size. However, it is important to note that if you select multiple faces that are not coplanar, such as faces that make up a fillet, the result will be a single face that is not a true representation of the intended feature, but rather a wavey version of the desired outcome. As shown in *Figure 2.16*, you can easily double-check the outcome of a **MERGE** feature using various inspection tools, such as **Zebra Analysis**. This tool shows how light would reflect off a given surface to display the changing curvature on a given surface. In *Figure 2.16*, you can easily see the difference between the stripes on the original CAD solid model on the left versus the stripes on the faceted version on the right, which utilize the **MERGE** feature to combine multiple faces into one around the fillet.

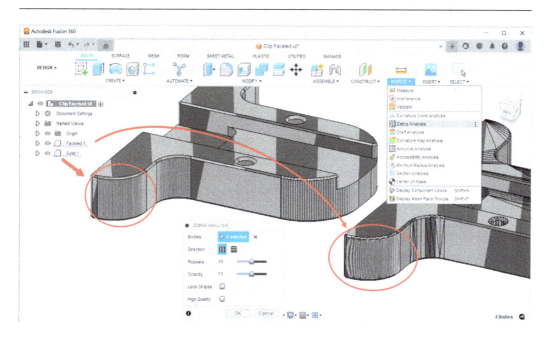

Figure 2.16 – Inspect merged faces using Zebra Analysis

The next feature we will highlight is **Fluid Volume**. This command, as well as the **Find Features** command, which we will cover next, is located in the **SOLID** tab | the **CREATE** panel, as shown in *Figure 2.17*.

Figure 2.17 – Find Features and Fluid Volume commands

You can create internal or external fluid volumes based on the type of solid body you are working with. *Figure 2.18* shows how you can create an external volume for a chess piece. After creating an external fluid volume, if you inspect your model with a section analysis, you will notice that the newly created external volume surrounds the chess pawn and does not interfere with the imported CAD body. Using this technique, you can quickly create a model that you can use to 3D-print a mold. You can use this newly created component as a starting point for creating the core. In this example, the desired outcome is a chess pawn. Therefore, we first create the cavity. Later, we can split this new component in half depending on how we will pull the mold apart. We can later add a pour basin and sprue before 3D-printing it. The GitHub repository for *Chapter 2* contains a CHESSPAWN.STEP file for you to try using this direct modeling feature. A finished version of the mold with two pulls with connector pins, as well as a pour basin and a sprue, is also provided (CHESSPAWNMOLD.STEP).

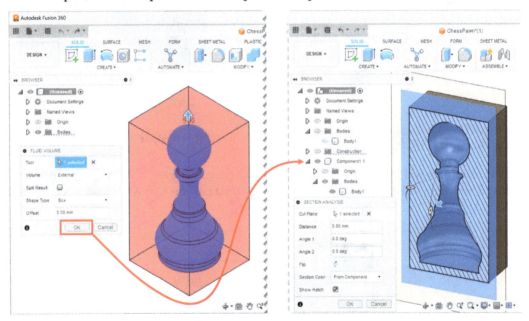

Figure 2.18 – Creating an external fluid volume for a component

In certain cases, we may have a CAD file for the mold itself and want to reverse-engineer the part and make changes so we can make sure it fits our requirements, such as size or aesthetics. In such cases, we could easily use the **Fluid Volume** tool and create an internal volume. *Figure 2.19* shows how you can create an internal fluid volume for a mold with two solid pulls and capping surfaces located at the top of the pour basin.

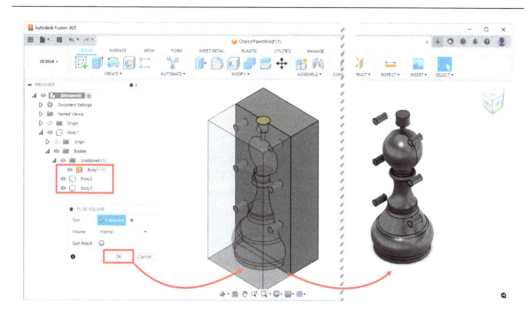

Figure 2.19 – Creating an internal fluid volume for a component

Notice that in this case, you will end up with one new component with seven bodies. The **Internal** volume option within the **Fluid Volume** command fills all possible cavities, including the tolerances between the pins and holes. If you do not need those extra solids, you can select, right-click, and delete them, as shown on the left half of *Figure 2.20*. Similarly, using the rectangle select tool, you can select all surfaces that make up the pour basin and sprue, right-click, and delete them, shown on the right-hand side of *Figure 2.20*.

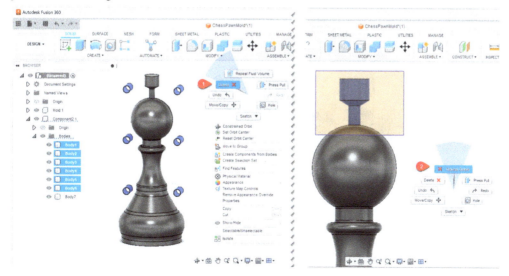

Figure 2.20 – Selecting and deleting bodies and faces

> **Important note**
>
> You may have noticed the similarity between the external fluid volume we can create with direct modeling and the simplify command we covered in the previous subsection related to parametric modeling. The underlying functionality between these two features is the same. However, as previously discussed, the **SIMPLIFY** feature is not entirely parametric and therefore could create issues when used in parametric modeling.

The next function, unique to direct modeling when working with imported CAD geometry, is one of my favorite features in Fusion 360. It is called **FIND FEATURES** and is in the **SOLID** tab | the **CREATE** panel. It allows you to select faces, bodies, or components and automatically recognize critical features, such as fillets, holes, chamfers, and patterns. After you use the find features command, and recognize a critical feature such as a fillet, it will show up in Fusion 360's browser with its designated name. You will be able to select the feature, right-click, and edit it. Depending on the position of the feature, you can easily change the dimensions of a given feature or completely delete it. We will cover how to use this command more effectively in the next section.

Direct modeling, as its name suggests, is not a parametric method for creating and editing geometry. Many of the modeling tools, such as fillet, hole, and pattern, have different behaviors during creation versus editing. Any feature you add to a solid model utilizes the exact same user interface as its counterpart from parametric modeling. The Fillet feature is a great example of this. In *Figure 2.21*, we are showing the application of a fillet to multiple edges.

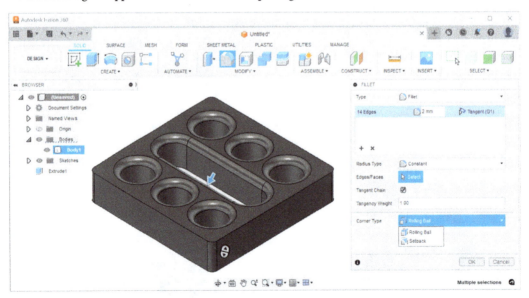

Figure 2.21 – Creating the Fillet feature in Direct modeling

Once the feature is added, we can edit the feature. However, as you can see in *Figure 2.22*, the **EDIT FILLET** dialog is very different and has a limited set of options compared to the creation dialog. This difference in creation versus editing is both the superpower of direct modeling and its Achilles heel, depending on how you want to model versus how you want to edit your models.

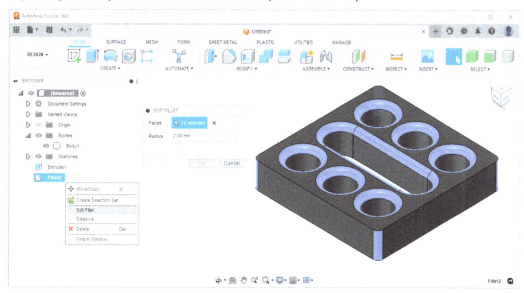

Figure 2.22 – Editing a Fillet feature in Direct modeling

The final group of features I want to highlight for direct modeling is related to working with components in an assembly context. In assembly modeling, we want to copy a given component and paste it to make duplicates so we can 3D-print multiples of the same part. We can duplicate components using one of many methods. We can select the component by right-clicking on it and choosing the **Copy** command, then repeating the process and choosing the **Paste** command. We can use one of multiple **Pattern** tools to create numerous instances of the selected component. Alternatively, we can simply insert a given component multiple times into our existing design from the data panel. If we are working with direct modeling, we can make a duplicated component an independent one. To accomplish this, select one of the duplicates, right-click on it, and choose **Make Independent**, as shown in *Figure 2.23*.

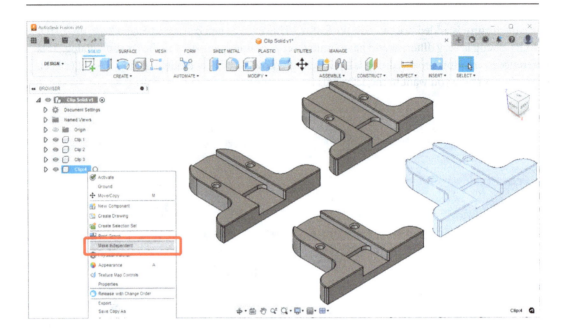

Figure 2.23 – Making a duplicated component independent from its parent

By breaking its link from the source component and making it independent, any changes we make to the original component will not propagate down to the duplicate. Similarly, any changes we make to the duplicated component will not impact the original component.

In parametric modeling, if you want to achieve similar results, you can only utilize copy component paste new component option. Or you can use the **Break Link** option after inserting a component into a design, as shown in *Figure 2.24*. There is no method for making a component independent after duplicating it using one of the three patterning methods located within the **SOLID** tab | the **CREATE** panel | the **PATTERN** pull-out menu (**Rectangular Pattern**, **Pattern on a path**, or **Circular Pattern**):

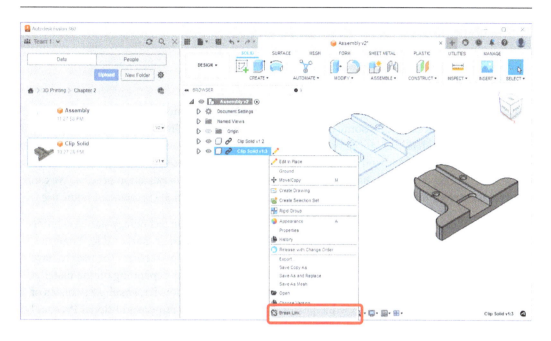

Figure 2.24 – Breaking the link of an inserted component

In this section, we covered how to work with direct modeling as well as parametric modeling. We highlighted various features that are unique to parametric modeling as well as direct modeling. We also highlighted some of the similarities between these unique features and their counterparts. In the next section, we will take a closer look at recognizing CAD features and editing them, which is a unique feature of direct modeling, as was introduced in this section.

Recognizing CAD features and editing them

In the previous section, we briefly touched on the find features command, which is unique to direct modeling. In this section, we will dive a little deeper into this feature and highlight its strengths and shortcomings. We will also utilize this feature with an example dataset in order to edit a CAD file we import with DFAM principles in mind.

As 3D printing has become more mainstream, we need categories to refer to a certain group of 3D printers based on their category. At the time of writing, there are seven categories of additive manufacturing:

- Material extrusion
- VAT photopolymerization
- Binder jetting
- Material jetting

- Powder bed fusion

- Direct Energy deposition

- Sheet lamination

There are multiple material extrusion methods, but probably the most common one is **Fused Filament Fabrication** (**FFF**). As some of the early patents for 3D printing started expiring after 2002, FFF technology has been going through democratization. Nowadays, anyone can purchase an FFF 3D printer for home use on common shopping websites such as www.amazon.com. VAT photopolymerization technology has also been through a similar democratization process. We can now purchase **stereolithography** (**SLA**)/**digital light processing** (**DLP**) 3D printers for home use.

However, even though some of the early patents around powder bed fusion technology started expiring around 2016, powder bed fusion 3D printers are not being manufactured or marketed for home use. This will likely continue to be the case as powder bed fusion technology requires the user to have a larger operating space with better ventilation and fire safety measures depending on the material being printed. In addition, as powder bed fusion involves the use of powder, small particulates of the material could be a safety hazard while postprocessing and would require additional **Personal Protection Equipment(PPE)**.

Regardless of the type of hardware we choose to 3D-print with, there are several guidelines and design rules we have to follow if we want to additively manufacture our designs. These rules are not set in stone and change based on various parameters. They can change based on the type of 3D printer we will use, the type of material we will print with, or even the manufacturer or the specific printer. To effectively utilize the information you learn in this chapter, you will need to know what 3D printing category/ printer you are designing for. There are several resources that list commonly accepted design rules for 3D printing. Hubs.com is one such resource you can utilize. *Figure 2.25* shows the published design rules for 3D printing on https://www.hubs.com/get/3d-printing-design-rules/.

DESIGN RULES
FOR 3D PRINTING

	Supported walls	Unsupported walls	Support & overhangs	Embossed & engraved details	Horizontal bridges	Holes	Connecting /moving parts	Escape holes	Minimum features	Pin diameter	Tolerance
	Walls that are connected to the rest of the print on at least two sides.	Unsupported walls are connected to the rest of the print on less than two sides.	The maximum angle a wall can be printed at without requiring support.	Features on the model that are raised or recessed below the model surface.	The span a technology can print without the need for support.	The minimum diameter a technology can successfully print a hole.	The recommended clearance between two moving or connecting parts.	The minimum diameter of escape holes to allow for the removal of build material.	The recommended minimum size of a feature to ensure it will not fail to print.	The minimum diameter a pin can be printed at.	The expected tolerance (dimensional accuracy) of a specific technology.
Fused deposition modeling	0.8 mm	0.8 mm	45°	0.6 mm wide & 2 mm high	10 mm	Ø2 mm	0.5 mm		2 mm	3 mm	±0.5% (lower limit ±0.5 mm)
Stereo-lithography	0.5 mm	1 mm	support always required	0.4 mm wide & high		Ø0.5 mm	0.5 mm	4 mm	0.2 mm	0.5 mm	±0.5% (lower limit ±0.15 mm)
Selective laser sintering	0.7 mm			1 mm wide & high		Ø1.5 mm	0.3 mm for moving parts & 0.1 mm for connections	5 mm	0.8 mm	0.8 mm	±0.3% (lower limit ±0.3 mm)
Material jetting	1 mm	1 mm	support always required	0.5 mm wide & high		Ø0.5 mm	0.2 mm		0.5 mm	0.5 mm	±0.1 mm
Binder jetting	2 mm	3 mm		0.5 mm wide & high		Ø1.5 mm		5 mm	2 mm	2 mm	±0.2 mm for metal & ±0.3 mm for sand
Direct metal Laser sintering	0.4 mm	0.5 mm	support always required	0.1 mm wide & high	2 mm	Ø1.5 mm		5 mm	0.6 mm	1 mm	±0.1 mm

Figure 2.25 – A set of design rules for 3D printing

As we can see in *Figure 2.25*, there are different requirements for different technologies. For example, if we were designing a part to be printed using selective laser sintering, we would not need to care about overhang angles. This category of 3D printers can print objects in any orientation without having to worry about creating support structures, as the part being 3D-printed is sufficiently supported by the powder below it. However, we need support structures to be generated for 3D printing with an FFF printer when we have features with overhang angles that are less than 45 degrees to the horizontal plane. *Figure 2.26* demonstrates how we can create different styles of support structures within Fusion 360 based on the overhang angles when we are in the manufacture workspace. We will cover this topic in *Chapter 10* in more detail.

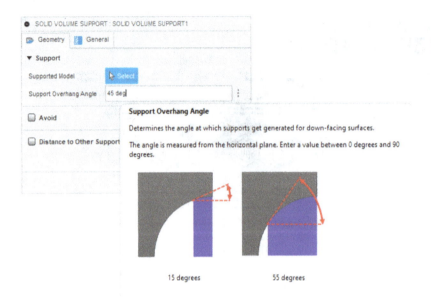

Figure 2.26 – Support structures for down-facing surfaces based on the overhang angle

Design rules such as embossed and engraved details, the minimum diameter of holes, escape hole requirements, and expected tolerances play an important role in how we design and prepare our models for 3D printing. If we are creating our own models from scratch, we can pay attention to how we design them to try and avoid issues when 3D printing them. However, if we are working with models created by other designers, and we are simply importing them to Fusion 360, we will need to inspect how the model was created and edit them as needed for 3D printing with additive manufacturing principles in mind.

Now, let's go through an example of opening a CAD file and inspecting it with native tools within Fusion 360. Later in the example, let's have Fusion 360 recognize and remove some small features not suitable for 3D printing and add new features for customizing the design.

We will start our exercise by opening the Chapter2-Example1.STEP file. As we previously covered in *Chapter 1*, you can open CAD files in Fusion 360 using the **File** | **Open** menu or upload the files to your desired Fusion team/project/folder, and then double-click it to open it in Fusions 360's canvas, as shown in *Figure 2.27*.

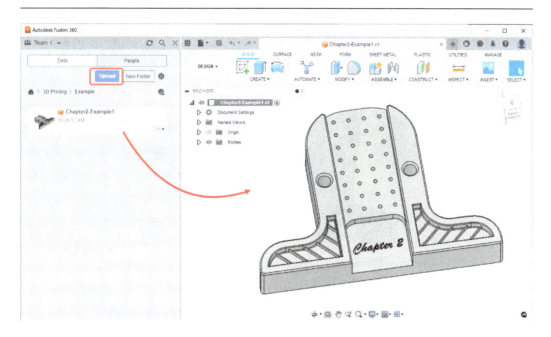

Figure 2.27 – Example file uploaded and displayed on Fusion 360's canvas

Now that we have the model in Fusion 360, let's start by visually inspecting it. This model has 2 large holes, 32 small holes, 2 pockets with webbing, engraved text, and many small fillets and chamfers. It also looks like the 32 small holes are in a pattern, but it is difficult to tell for sure at this point in time.

Let's start using some of the feature recognition tools in Fusion 360's direct modeling environment and focus on holes in this design. As shown in *Figure 2.28*, after initiating the Find features command, selecting **BODY1** as the input, and unchecking all but the Hole and Rectangular pattern feature, we will find the 2 large holes and the 32 small ones as 2 separate rectangular patterns. Looks like our visual assumption was correct that there was a pattern to the small holes, but it was not a single continuous pattern – rather 2 different patterns.

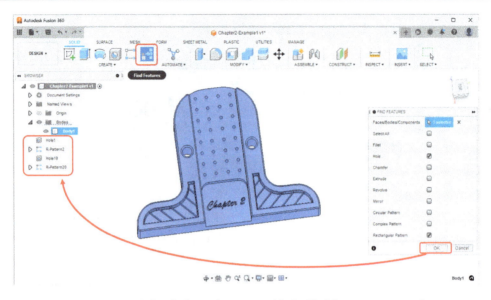

Figure 2.28 – Finding holes and patterns with the Find features command

Now, let's investigate the dimensions of the holes. If we select **Hole1** from the browser, right-click, and select **Edit Hole**, as shown in *Figure 2.29*, we will be able to inspect and change, if needed, all the dimensions of this hole. As the dimensions listed for this hole are above the minimum dimensions for a printable hole for FFF, we will not make any changes. We can simply click **Cancel** and start focusing on the small holes.

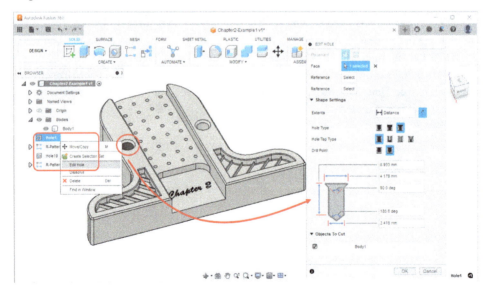

Figure 2.29 – Edit Hole1 and view its dimensions

Now, let's look at one of the smaller holes. By selecting an edge of one of the small holes, we can quickly see its diameter information in the bottom-right corner of the canvas, as shown in *Figure 2.30*.

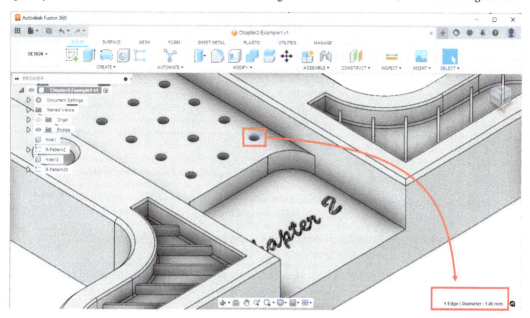

Figure 2.30 – Selecting a circular edge to view its diameter

This investigation tells us that the diameter of all the small holes is 1 mm, which is smaller than what our FFF printer could possibly print. So, we can simply delete these holes and drill holes once our 3D print is complete if they are absolutely necessary. To delete all 32 holes, we have several options. We can orient our view so we are looking at the model from View cube = TOP, select all the faces with rectangle select, and right-click and delete. This method was demonstrated in the right half of *Figure 2.20*. Alternatively, as shown in *Figure 2.31*, we can select the two rectangular patterns from the tree, right-click, and delete them, which will delete all the holes that belong to these patterns.

Figure 2.31 – Selecting and deleting rectangular patterns within the browser

The next feature we may want to focus on is the engraving. Looks like there is embossed text within this file. There are various ways to remove embossed text. Similar to how we selected faces in the past and deleted them, we can remove all the faces related to the emboss feature. However, selecting multiple complex faces such as the one in this example may not be straightforward. In such cases, we may want to use a different approach. Let's start by creating a new patch by going to the **SURFACE** tab and accessing the patch command within the create panel. We will select the six edges that we want our patch to cover. This action, as shown in *Figure 2.32*, will create a new surface body.

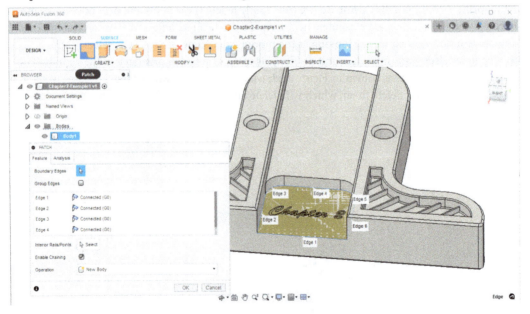

Figure 2.32 – Using the patch command to create a new surface

Once we have a new surface body created, we can utilize the boundary fill command, which is accessible in both the **SURFACE** and **SOLID** tabs under the create panel. Using the boundary fill command, as shown in *Figure 2.33*, we can select our solid and surface body as tools and rectangle select all the cells we wish to encapsulate.

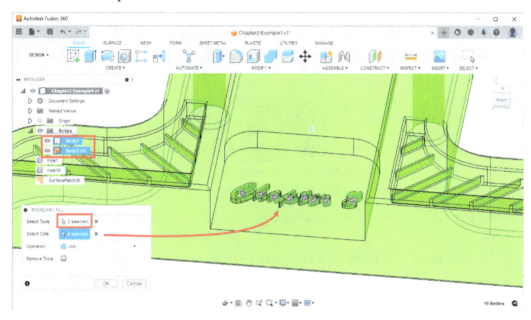

Figure 2.33 – Using the Boundary fill command to remove the engraving

The boundary fill command is similar to the fluid volume command in that it creates new solids. But this command also allows us to join those new solid bodies to our existing solid body. Once the boundary fill command is complete, we will have eliminated the engraving. We no longer need the surface patch or surface body we created, which means we can select and delete it from the browser, as shown in *Figure 2.34*.

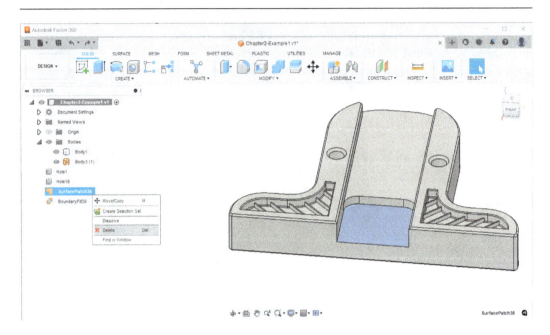

Figure 2.34 – Deleting the surface patch, which is no longer needed

Next, we can move on to detecting other features in our model. Let's go back to the find features command, select the body, and find all chamfers. By systematically selecting the desired features to find, we can modify or remove them without impacting other features within the model. As shown in *Figure 2.35*, after selecting the chamfer within the Find Features command, Fusion 360 will find a single chamfer feature associated with the model. We can simply select and delete it from the browser to eliminate this small feature.

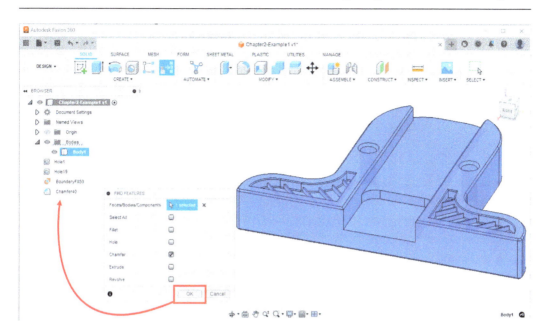

Figure 2.35 – Using the find features command to find chamfers

Next, we can repeat the same process and find all the fillets within the same body. Once Fusion detects all the fillets, we can click through them within the browser to identify which fillet they are on the canvas. In this case, **Fillet42** represents the small fillets within the slot/webbing zone. We can simply select **Fillet42** from the browser, as shown in *Figure 2.36*, right-click, and delete it if its dimensions are below what our 3D printer can manage.

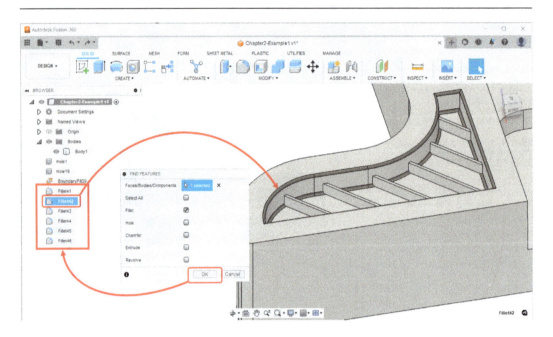

Figure 2.36 – Detecting fillet features using the find features command

In this section, we introduced the concept of design rules for 3D printing and how they can change based on 3D printing technology, material, and even the specific printer. Then, we used Fusion 360's direct modeling tools on imported CAD geometry to detect, inspect, edit, and delete features that were not fit for our needs.

It is important to remember that Fusion 360 is a full-featured CAD package, which allows you to add new features to imported CAD geometry. For example, if you wanted to add new pockets, holes, fillets, or chamfers, you would be able to simply add them using the functionality available within the **CREATE** and **MODIFY** panels within the **SOLID** tab.

In the next section, we will work with mesh files in a similar manner and edit them directly with Fusion 360 to prepare them for 3D printing.

Working with Mesh files natively in Fusion 360

In the previous section, we talked about how to open CAD files and edit them for 3D printing. In this section, we will insert Mesh files and learn how to work with them in various ways.

After inserting a Mesh file into Fusion 360, we have three ways we can work with those models. We can edit inserted mesh bodies in the following ways:

- As Mesh bodies

- After converting them into solid bodies

- By recreating them as solid bodies

Editing inserted Mesh bodies as mesh bodies

The mesh functionality within Fusion 360 is available for both parametric and direct modeling methodologies. When working with mesh files, there are several reasons why we may want to simply keep them as mesh bodies to prepare them for 3D printing purposes. Using a mesh body, we can do quick manipulations, such as scaling a mesh body up or down with **SCALE MESH**, cutting a mesh body in two with **PLANE CUT**, hollowing a mesh body with **SHELL**, and combining or removing a mesh body with other mesh bodies with **COMBINE**.

All these operations are available within the **MESH** tab | the **MODIFY** panel, as shown in *Figure 2.37*.

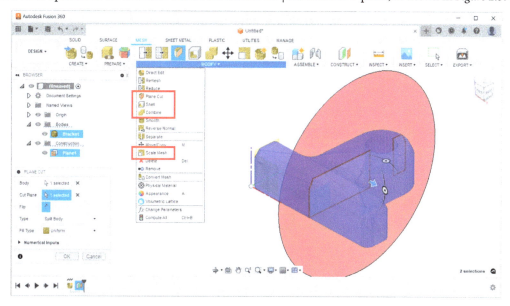

Figure 2.37 – Editing a mesh body in Fusion 360 with the Plane Cut feature

As mesh bodies are native body types to Fusion 360, you can also utilize mesh bodies as inputs for volumetric latticing if you have access to the Design extension. We will cover various options we have for latticing in more detail in *Chapter 4*.

If you select the mesh body and make it a component, as shown in *Figure 2.38*, you can perform additional print preparation activities, such as mirroring components that have mesh bodies in them.

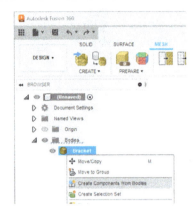

Figure 2.38 – Creating a component from a mesh body

Once a mesh body is included within a component, you can create various patterns using that component as the input object type, as shown in *Figure 2.39*.

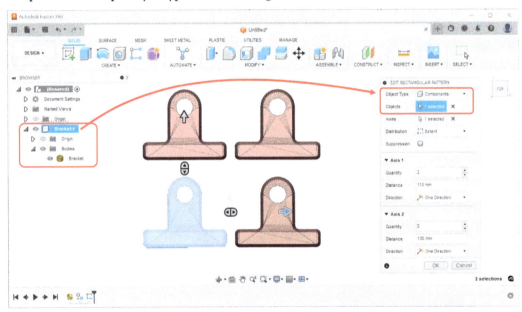

Figure 2.39 – Patterning components with mesh bodies

Unlike the **SCALE**, **PLANE CUT**, and **SHELL** commands, the **COMBINE** command within the **MESH** tab | the **MODIFY** panel requires multiple body inputs. The **COMBINE** command allows you to join, cut, merge, or intersect multiple mesh bodies. *Figure 2.40* shows how you can create a new hole feature within an inserted mesh body in three steps. The first step is to create a cylinder using solid modeling techniques. In the second step, we convert that cylinder into a mesh body using the **tessellate** command, as shown in the screenshot in the following figure. The final step is to use the **combine** command with the cut operation to remove the tessellated cylinder from the bracket.

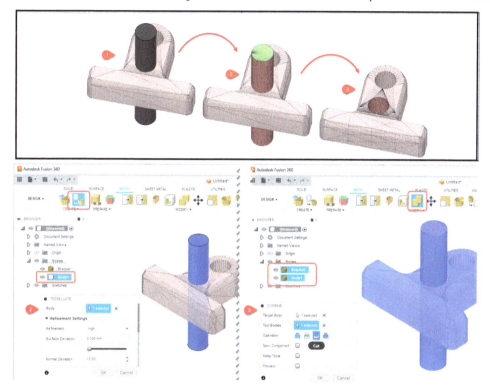

Figure 2.40 – Using the combine-cut feature on a mesh body to create a hole

Editing inserted Mesh bodies after converting them into solid bodies

Even though working with mesh bodies in their native state is a quick way to manipulate the entire body and add or remove certain features, we cannot create all the features necessary unless we are working with solid bodies. In such cases, we may need to convert a mesh body into a solid body.

When automatically converting mesh bodies into solid bodies, Fusion 360 offers three options:

- Faceted (available to all users)
- Prismatic (*N/A for Personal Use*)
- Organic (*N/A for Personal Use* and *Requires Access to the Design extension*)

In this chapter, we will only highlight the Faceted and Prismatic options. The Organic option will be highlighted in more detail in *Chapter 3* after we generate an organic mesh body using the simulation capabilities of Fusion 360.

The **FACETED** method for converting mesh bodies into solid bodies will simply change the body type from mesh into solid. In *Figure 2.41*, we can see the steps we need to take to convert a mesh body into a faceted solid body. Once the conversion is complete, we can select any of the planar faces on the solid body and delete them using either the keyboard *Delete* key or the delete command, which we can access by right-clicking or from our shortcuts, which we can reach by pressing *S* on the keyboard to search for them. *Figure 2.41* also shows how we can quickly convert all the planar faces of a given object into clean surfaces and later use those surfaces for solid modeling operations, such as adding additional bodies or creating holes. However, the faceted option is not capable of automatically converting nonplanar shapes into clean surfaces. To accomplish that, you will need the second option, which is **PRISMATIC**.

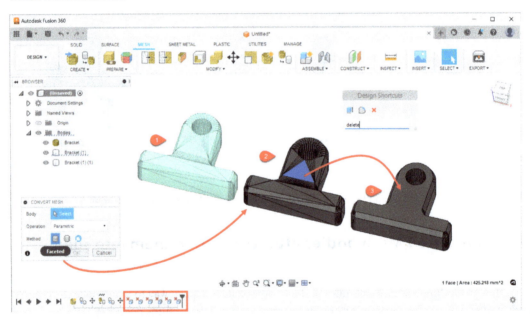

Figure 2.41 – Convert Mesh bodies into faceted solid bodies and delete planar faces

In order to use the prismatic method to convert mesh bodies into solid bodies, Fusion 360 has one key prerequisite. You will need to define face groups prior to using the prismatic conversion. Fusion 360 offers automatic face group detection under the **MESH** tab | the **PREPARE** panel. Using the Fast or the Accurate method, and if necessary, editing the settings of these methods, you can have Fusion 360 display each face group with a unique color by utilizing the Display mesh face groups command, located under the **INSPECT** panel. The Accurate method often gives the best outcome, as shown in *Figure 2.42*.

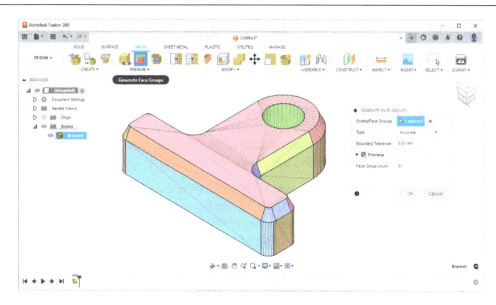

Figure 2.42 – Preview of the outcome of the Generate Face Group command

In some cases, you may need to do additional work to achieve the desired face group outcome. You can start by combining face groups in parametric mode or accessing the Direct edit mode to create face groups manually. Once you have all the face groups needed, you can utilize the **PRISMATIC** conversion method, as shown in *Figure 2.43*, and convert your mesh body into a solid body. At this point, you are ready to add any new features, including complex shapes such as fillets and chamfers.

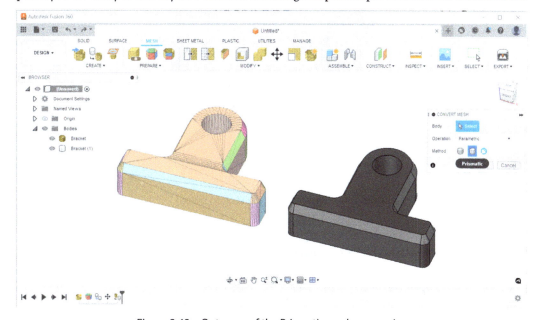

Figure 2.43 – Outcome of the Prismatic mesh conversion

Editing inserted Mesh bodies by recreating them as solid bodies

Even though converting mesh bodies into solid bodies is the best way to make sure we can edit them with all the solid modification functions included in Fusion 360, sometimes for very complex models, the mesh conversion may not be successful. In such cases, we can use the inserted mesh body as a reference and create sketches. Later, we can utilize profiles within those sketches as input for the **EXTRUDE, REVOLVE, SWEEP,** or **LOFT** commands to create the desired solid outcome. This is by far the most involved method for converting mesh bodies into solids, but it is a powerful tool in our toolset. Let's demonstrate this workflow using the same dataset.

After inserting our mesh and moving it to ground, we can utilize the **CREATE MESH SECTION SKETCH** command located in the **MESH** tab | the **CREATE** panel. After selecting the body, Fusion automatically selects a section plane. This may or may not be the section plane we need to create our sketch. In this case, we want to create a sketch parallel to the XY plane. We can hit the select button next to **Section Plane**. This will display the global planes and the global axes. We will select the XY plane and, using the on-screen manipulators, move it up by 10 millimeters. This will give us a preview of the mesh section sketch, as shown in *Figure 2.44*.

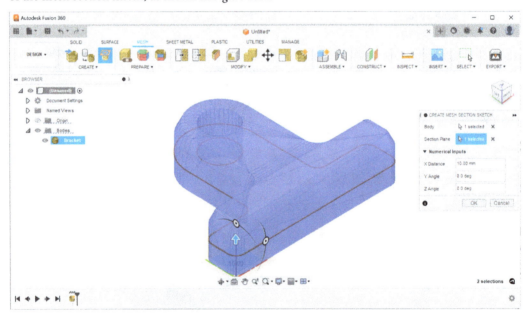

Figure 2.44 – Preview of mesh section sketch

Next, we can select Sketch 1 in the browser, right-click, and edit it. In this example, we will utilize the **FIT CURVES TO MESH SECTION** command located in the **SKETCH** tab | the **CREATE** panel. Using the various **Fit Curve Type** options, we can select a segment of the mesh section sketch and create mesh lines, arcs, circles, and splines, as shown in *Figure 2.45*.

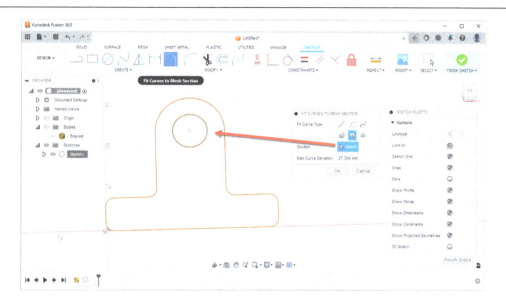

Figure 2.45 – Creating a circular profile using the FIT CURVES TO MESH SECTION command

Once we are done creating our sketch geometry, we can use the extrude command with a two-sided direction to thicken this profile. Prior to extruding the profile, we can utilize the measure command, located within the **INSPECT** panel, to figure out how thick this part should be. In this example, our part is a total of 25 millimeters, so we will extrude it 15 millimeters in one direction and 10 millimeters in the other, as shown in *Figure 2.46*.

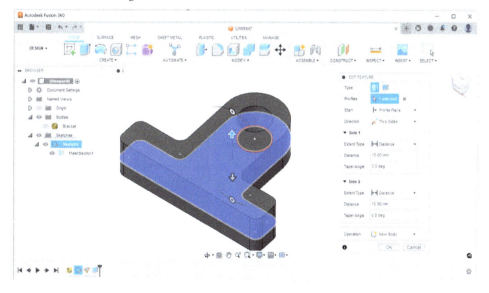

Figure 2.46 – Using the extrude command to turn a sketch profile into a solid body

Next, we can start focusing on the chamfer feature, which should be replicated around the top edge of this part. We can start by utilizing the **MEASURE** command, located in the **INSPECT** panel, to understand the XYZ delta that makes up the chamfer feature we are trying to replicate. As we can see in *Figure 2.47*, this part has a chamfer that is two-sided. In one direction, it is 3 millimeters and in the other, it is 6 millimeters.

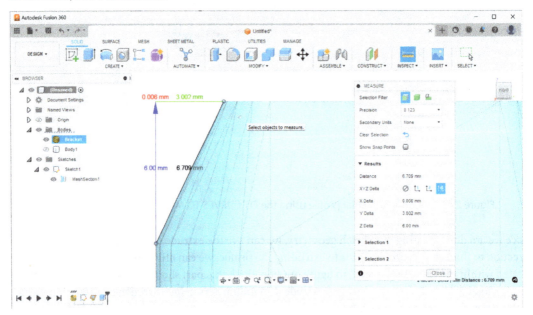

Figure 2.47 – Measuring the mesh to dimension the chamfer

Now that we know the dimensions of the chamfer that the mesh body had, we can add a chamfer feature to the solid body. Simply activate the **CHAMFER** feature located under the **MODIFY** panel, select all 14 edges around the perimeter of the top surface, change the chamfer type to **TWO DISTANCE**, and enter the corresponding values into the dialog, as shown in *Figure 2.48*.

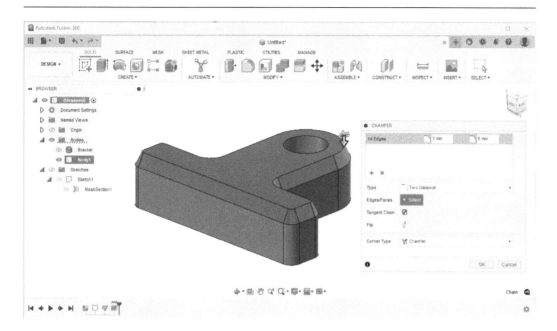

Figure 2.48 – Using the chamfer feature to add a two-distance chamfer to the perimeter

Now that we have created a mesh section sketch, utilized that to create sketch profiles, extruded those profiles, and added the necessary chamfer, we have fully reverse-engineered this part, which started as an inserted mesh and ended as a solid body. This workflow required more steps than the previous workflows we highlighted, but it can be very handy if the parts we are trying to reverse-engineer cannot be recreated using other methods.

Summary

In this chapter, we went over how to edit CAD/Mesh models with various **Design for Additive Manufacturing (DFAM)** principles in mind. We highlighted how to turn on/off the design history. We demonstrated the differences in the workflow when using parametric versus direct modeling. We then learned how to automatically detect features for imported CAD models. We also went over how to edit both CAD and mesh bodies based on DFAM principles. We ended the chapter by going over several methods to convert mesh files into solid bodies that can be further modified for manufacturing.

In the next chapter, we will cover several techniques we can use within Fusion 360's Design workspace with Automated modeling and Simulation workspace with Shape optimization for lightweighting parts with DFAM principles in mind. We will also learn how to utilize simulation to identify potential performance failures and make design changes to account for them.

3

Creating Lightweight Parts, and Identifying and Fixing Potential Failures with Simulation

Welcome to *Chapter 3*. One of the main benefits of additive manufacturing is the ability to create lightweight parts. In this chapter, we will cover various tools we have within Fusion 360 for lightweighting parts and creating organic-looking models that we can 3D print. We will start by highlighting how to utilize the automated modeling capabilities within the design workspace to generate design alternatives. We will discuss the differences between automated modeling and generative design. Next, we will utilize shape optimization within the simulation workspace to create lightweight and structurally efficient components. We will conclude the chapter by showcasing how to simulate the components we create to check, detect, and fix failures common in 3D-printed components.

In this chapter, we will cover the following topics:

- Getting started with Automated modeling
- Shape optimization
- Structural simulation to detect and fix common failures

By the end of this chapter, you will have learned how to use automated modeling to create organic shapes. You will be able to use shape optimization to create structurally efficient components. You will also know how to use linear static stress to detect stress concentrations to eliminate them by making informed design changes.

Technical requirements

All the topics covered in this chapter are accessible to Startup and Educational Fusion 360 license types. If you have a Commercial Fusion 360 license, in order to conduct a shape optimization study, you will also need access to the Simulation extension (`https://www.autodesk.com/products/Fusion-360/simulation-extension`), or you can use Autodesk Flex tokens to solve shape optimization studies (`https://www.autodesk.com/benefits/flex`).

Even though we will not be conducting a generative design study in this chapter, if you have a Commercial Fusion 360 license and would like to conduct a generative design study, you will need access to the Simulation extension (`https://www.autodesk.com/products/Fusion-360/simulation-extension`), or you can use Autodesk Flex tokens to solve generative design studies (`https://www.autodesk.com/buying/flex?term=1-YEAR&tab=flex`).

Autodesk offers a 14-day trial for the Simulation extension for all commercial users. You can activate your trial within the Extensions dialog of Fusion 360.

Neither Automated Modeling, Generative Design, nor any of the simulations mentioned previously (Shape Optimization and Linear Static Stress) are available for Personal/hobby licenses of Fusion 360.

The lesson files for this chapter can be found here: `https://github.com/PacktPublishing/3D-Printing-with-Fusion-360`.

Getting Started with Automated modeling

In previous chapters, we covered how to use parametric modeling and direct modeling in order to make changes to existing designs. Using solid and mesh modeling techniques, we created holes, pockets, and hollowed-out parts, which are all useful methods for lightweighting parts.

The Automated Modeling functionality within the Design workspace of Fusion 360 is a great tool for designers to create organic-looking lightweight components. In this section, we will focus on creating new models using this generative technique. After choosing a design alternative, we will modify the volume of the outcome. Later, we will manually edit the organic shape to further lightweight the design.

Automated modeling is available within the **DESIGN** workspace's **SOLID** tab's **AUTOMATE** panel. Using automated modeling, users can generate design alternatives between geometries that they want to connect. This function requires a minimum of two faces to be selected as input faces. It also allows users to select zones where a model should not be created. When using this command, Fusion 360 will generate six design alternatives. It will then allow you to explore the design that best suits your needs. Three of the design alternatives are created using smooth connections between the selected faces. The remaining three are created using sharp connections. Now, let us explore this functionality with an example dataset and walk through the iterative process a designer may follow while working on a project.

Figure 3.1 shows a microphone assembly and a stand. This model is made up of several bodies with individual names, and we will refer to them by those names throughout this section.

Let's imagine we are a designer who is tasked with coming up with a new version of the **Connection-left** and **Connection-right** bodies for this microphone stand. Those bodies were originally created using the **LOFT** command within Fusion 360, and they are shown in green in *Figure 3.1*. These connection arms connect the base to **Mount-right** and **Mount left**, which are represented in yellow in the same figure:

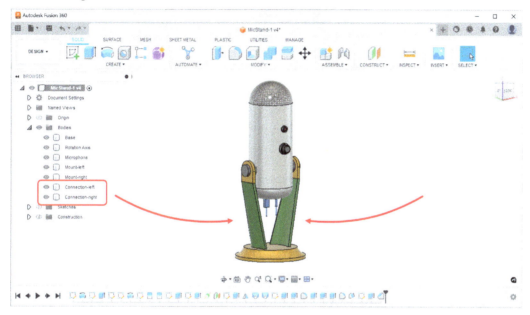

Figure 3.1 – Microphone stand design with unique bodies for connection arms

Using existing bodies that are present in this Fusion 360 design, we can start this exploration by turning off the visibility of the bodies named **Connection-left** and **Connection-right**. Then, we can access the **Automated Modeling** command within the **AUTOMATE** panel.

Next, we can select the four faces we wish to connect to create a new geometry and select **Generate Shapes**, as shown in *Figure 3.2*. Note that this figure was captured by turning the opacity of all visible bodies down to 50% to better display the four selected faces.

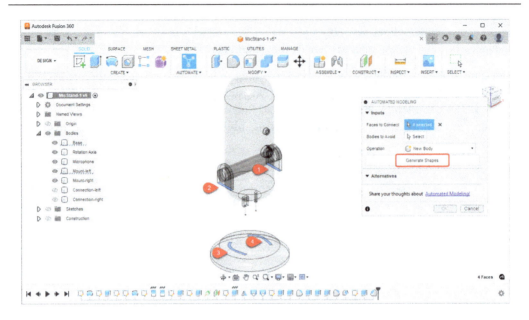

Figure 3.2 – Selecting the faces to connect within Automated Modeling

Once all the design alternatives are generated, they will have a green checkmark next to them and we can explore the outcomes. As shown in *Figure 3.3*, the **AUTOMATED MODELING** dialog displays six design alternatives. As we select each alternative, a preview of that geometry will be displayed on the canvas.

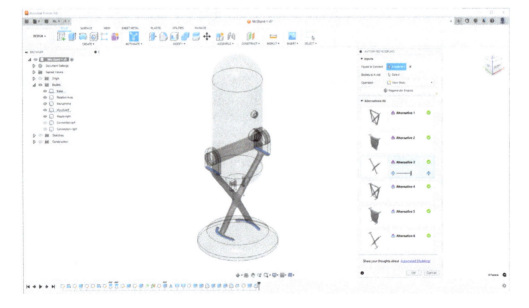

Figure 3.3 – Automated modeling generates six design alternatives

If you look closely, you will notice that the first three alternatives have an icon with a purple form body and a transparent object underneath it. The following three design alternatives have a purple form body, a blue solid body, and a transparent object underneath it. A close-up of the icons for the first three alternatives versus the next three can be seen in *Figure 3.4*. This subtle difference in the iconography refers to the connection type of the generated design alternative. Alternatives without the gray solid body create smooth connections. Alternatives with the gray solid body create sharp connections.

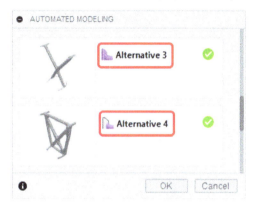

Figure 3.4 – Different icons represent a change in the connection types for design alternatives

Each design alternative also offers a volume slider. Using this volume slider, we can generate a preview of the outcome with varying volumes. *Figure 3.5* shows the difference between choosing a minimum volume (left) versus a maximum volume (right) for the **Alternative 2** design.

Figure 3.5 – Differences in volume after modifying the volume slider

Investigating all the design alternatives, let's say that we liked **Alternative 6** the best. After adjusting its volume slider toward the minimum, we can press **OK** and generate the outcome as a new body. This new body will be added to the tree. *Figure 3.6* shows that a new body named **Body30** has been added to the list of bodies and a new group called **AutomatedModeling3** has been added to the timeline:

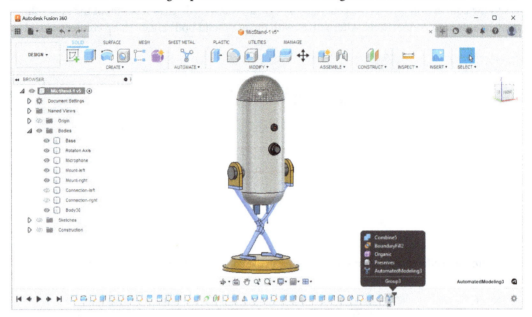

Figure 3.6 – Selecting Alternative 6 and generating a body

At first glance, the outcome looks promising. However, after a quick inspection, we can see that we have made a mistake and that the outcome is penetrating into where the power and headphone wires would be connected to the microphone. To get a precise answer, we can utilize the interference detection tool located in the **INSPECT** panel, as shown in *Figure 3.7*. After selecting the bodies named **Microphone** and **Body30** and computing the interference between them, we can see that there is more than 1,000 mm^3 of interference between these two bodies:

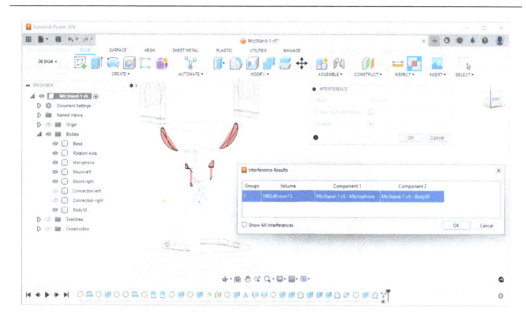

Figure 3.7 – The interference between the microphone and the organic connection arm

To address this interference and to create shapes that do not interfere with the microphone, let's go ahead and suppress the **AutomatedModeling3** feature from the timeline and create a new automated modeling feature. This time, in addition to selecting the four faces to connect, we will also select two bodies to avoid: **Rotation Axis** and **Microphone**, and select the **Opacity** checkbox so the bodies to avoid are transparent. The new setup for this automated modeling can be seen in *Figure 3.8*.

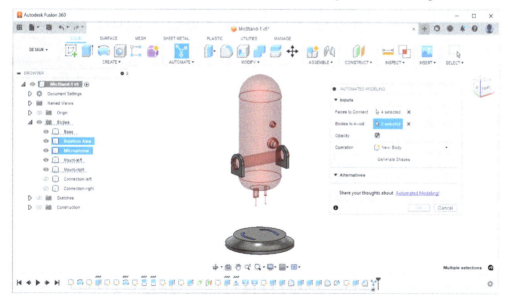

Figure 3.8 – A new automated modeling setup with a couple of bodies to avoid

After generating the new shapes, we now see six new design alternatives to choose from. And according to Fusion 360, they all avoid interfering with the microphone and the modeled cables. After inspecting the outcomes, let's choose **Alternative 4** and adjust its volume slider to the minimum and select **OK** to generate the new body, as shown in *Figure 3.9*:

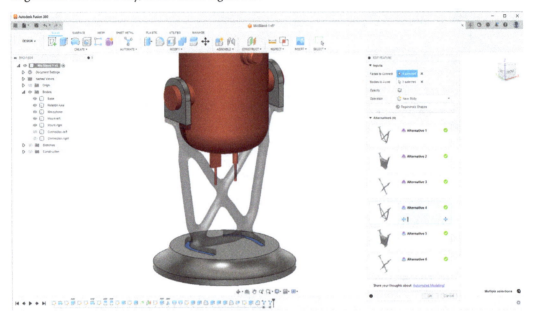

Figure 3.9 – Six new design alternatives that avoid interfering with the microphone

As we can see in *Figure 3.10*, we now have a new body named **Body46** in the **BROWSER** panel and we can see that a new automated modeling feature has been added to the timeline. Upon closer inspection, we can see that the new body is not interfering with the cables. However, this was done by removing the microphone and the cables from the newly generated shape. This new body theoretically meets our objectives; however, in practice, we need a much larger clearance between our cables and the connection arms so we can plug and unplug these wires:

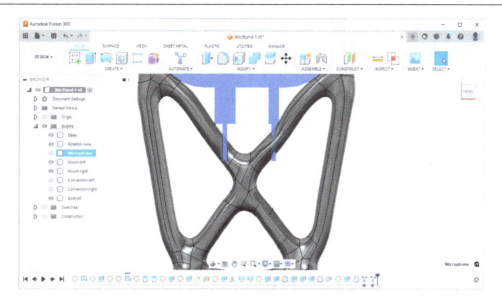

Figure 3.10 – Outcome of the second iteration of automated modeling

To achieve this new objective of a larger clearance between our microphone and the new connection arms, we need to create a new geometry. Fusion 360 can be utilized to create a new sketch along the *XZ* axis, and we can project the profile of the microphone and, if needed, offset it to create an additional tolerance zone. Later, we can use the profile we created and use the **REVOLVE** command to create a 130-degree revolution, as seen in *Figure 3.11*:

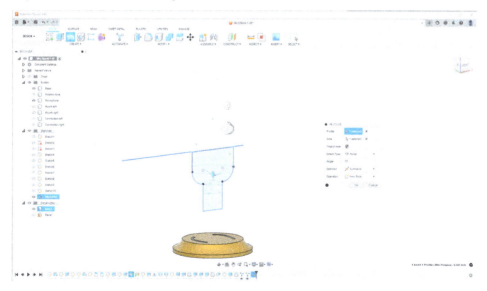

Figure 3.11 – Creating a new body using the REVOLVE command based on the profile of the microphone

Once we generate this new body, we can rename it. In this example, I renamed the new body `Avoid Zone`. Now, we're ready for our third and final automated modeling feature. Let's go ahead and suppress our prior automated modeling features and activate automated modeling one last time. After selecting the same four faces to connect, we can now select the newly created **Avoid Zone** as an input for the bodies to avoid. The new setup for the final automated modeling can be seen in *Figure 3.12*:

Figure 3.12 – Final setup for automated modeling with the newly created Avoid Zone

After generating the new shapes, we now see six new design alternatives to choose from. If we choose **Alternative 6** and utilize the minimum volume, we will end up with the model preview that we can see in *Figure 3.13*:

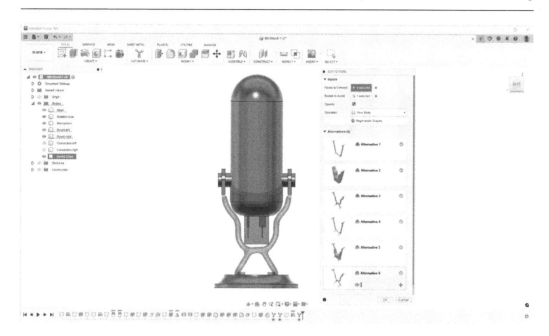

Figure 3.13 – Alternative 6 for the final automated modeling setup

It is always a good idea to take a closer look at the outcome and make sure it meets our design objectives and that the output geometry is geometrically pleasing. As we can see in *Figure 3.14*, Fusion 360 created a new body utilizing T splines to create the shapes, used the boundary fill command to convert surface shapes into solid objects, and finally, used the combined feature cut operation to remove avoid bodies from the output shape. This final action of combined cut created two notches in our otherwise organic-looking geometry. We can easily make a change to the combined feature and remove one of the four tool bodies (in this case, **Avoid Zone**) to eliminate these two notches.

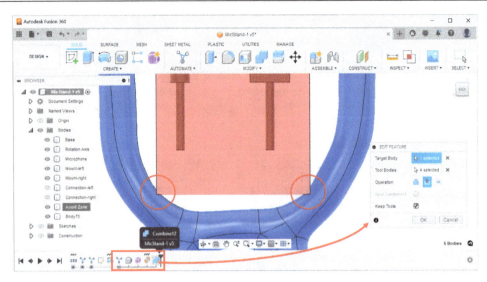

Figure 3.14 – Automated modeling creates a group of features that can be edited individually

With this final edit, we have created a design alternative that both fits our requirements and is drastically different compared to the original design. But is this design lightweight compared to the original? Even though we did not explicitly enter criteria around lightweighting the part as we were generating, we were consistently selecting the minimum volume for each selected design alternative. Let's take a closer look at the original part and its properties. As we can see in *Figure 3.15*, the component named **Original Connection** had a volume of 41,843 millimeters cubed. Note that to access the **PROPERTIES** dialog, we first created a new component and moved the bodies named **Connection-left** and **Connection-right** into this new component. Later, we right-clicked and selected **Properties** to access the **PROPERTIES** dialog.

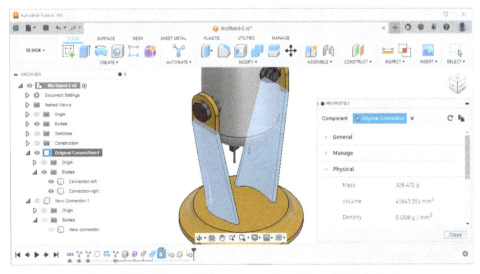

Figure 3.15 – Mass and Volume properties of the original design

If you repeat the same steps, create a new component, and move the new connection body into this new component, we can take a look at its properties and see that the volume for this new component is 16,591 millimeters cubed. This means that the new design is roughly 40% of the volume and mass of the original design. *Figure 3.16* shows the properties of the new connection body created using automated modeling.

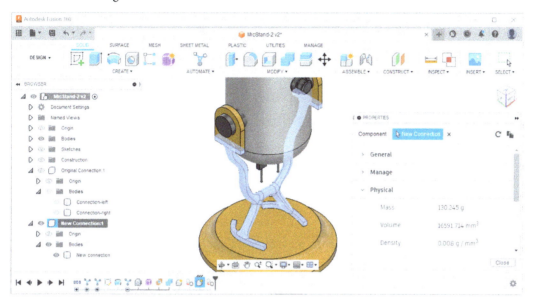

Figure 3.16 – Mass and Volume properties of the new design

Even though automated modeling gave us both an organic look and a lightweight design, we do not know for sure how this geometry will perform in a real-world setting. In order to verify that our model can withstand the loads associated with the weight of the microphone in various positions, we have to either build a prototype and test it or perform a simulation. Another tool we could have used to generate both organic and lightweight designs is generative design. Generative design is its own workspace within Fusion 360 and requires similar inputs for connecting faces and obstacle geometry. Using generative design, you also have the capability to enter loads and boundary conditions to simulate the forces this connection arm will have to bear. In addition, generative design allows users to select which manufacturing model they wish to produce the output geometry. Options such as **Additive** manufacturing, subtractive manufacturing (**2.5 axis**, **3 axis**, or **5 axis milling** machines), as well as **2 axis cutting** and **Die casting**, can be accounted for. Generative design also requires you to select which materials should be considered when creating designs that meet the requirements. As you can see in *Figure 3.17*, you can generate tens, if not hundreds, of outcomes based on the manufacturing method and material that meet certain design objectives.

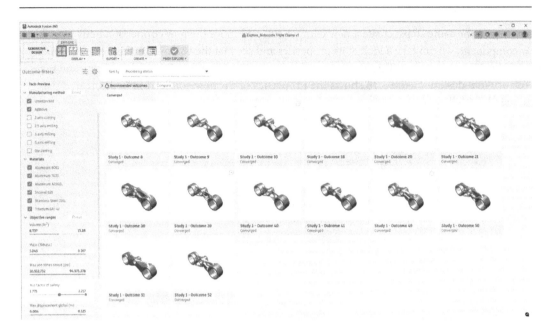

Figure 3.17 – Generative design outcomes of a motorcycle component that meets design criteria

Regardless of the method we choose to generate organic and lightweight geometries, it is always good practice to perform a final simulation to make sure that these computer-generated shapes meet our performance criteria.

In this section, we covered how to apply automated modeling to generate organic-looking shapes. We went through an iterative process of design utilizing automated modeling and updated our design criteria to meet a certain objective. We used the microphone stand as an example and compared the original design to the final design to ensure the outcome was a more lightweight version. In the next section, we will follow a similar example using the Shape Optimization simulation method to make sure that the design we create considers the performance criteria.

Utilizing Shape Optimization

In the previous section, we covered how to use automated modeling in order to generate an organic and lightweight design. In this section, we will introduce performance-based modeling and learn about how to utilize Shape Optimization for generating similar outcomes.

Shape Optimization is an analysis type available within the **SIMULATION** workspace of Fusion 360. Using Shape Optimization, engineers can create lightweight designs based on predetermined performance criteria. However, it is important to note that the outcome of a shape optimization study does not actually reflect the stresses created by the loads. Therefore, we should always validate our new design using additional simulations such as a static stress study. In this section, we will focus on creating new models using the shape optimization simulation technique. After choosing a design alternative, we will incorporate the outcome into our original design, both as a mesh body and as a solid body, and talk about the benefits and drawbacks of each method.

To conduct a Shape optimization study, at a minimum, we are required to select a single **Target Body** to safely remove material from. We also have to assign materials to that body. Finally, we need to apply constraints and structural loads that act on that body.

Loads can be applied by simply selecting faces and applying force or pressure. Alternatively, we can add loads to faces, edges, or corner points of other components that are in contact with our target body. As long as we define the contact conditions in our assembly properly, the target body will be optimized based on those loads. We can also select secondary bodies and have Fusion 360 automatically calculate their center of mass and apply a load based on their weight to our target body as remote loads.

Shape optimization also allows us to create preserved regions based on common shapes such as a box, a cylinder, or a sphere. **Preserved Region** allows us to designate a certain zone to not be optimized. For example, if we want to preserve a hole that will be used as a connection interface for a bolt, we can select it and designate it as a preserved region so that the shape optimization does not remove material from that segment of the model. Within shape optimization, we can also ask Fusion 360 to give us a symmetrical output. In such cases, we will have to define the a symmetry plane so that the outcomes adhere to this symmetry condition. The most important part of a shape optimization study is the optimization criteria. Fusion 360 allows us to maximize the stiffness while minimizing the mass of the target body. In certain simulations, minimizing the mass using shape optimization may result in a design with numerous slots and braces. In such cases, Fusion 360 also allows us to control the minimum member size of any webbed design output.

Now, let us explore this functionality with the same dataset and walk through an example process a designer may follow while working on this project using Shape optimization.

Figure 3.18 shows the base and the mounting brackets for the microphone stand we introduced in the previous section of this chapter in yellow. Using Fusion 360's sketching tools, we can create a rectangle at the bottom of the two mounting brackets and a circle on top of the base. Then, we can utilize the **LOFT** command and create a solid body between these two sketch profiles. This new body will be the basis for our target body for shape optimization. Let's rename this body `Shape to optimize`:

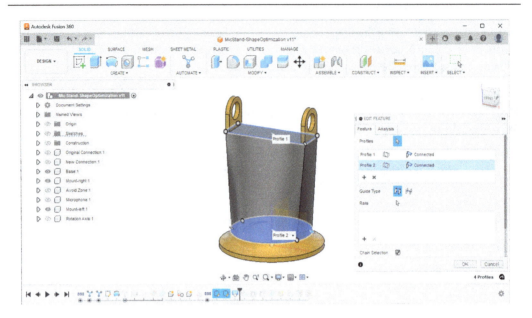

Figure 3.18 – Creation of a new body with the LOFT command
between the base and the mounting brackets

Next, we will utilize the body named **Avoid Zone**, which we created in the previous section of this chapter. Let's start by selecting the exterior faces of this body, as shown in blue in *Figure 3.19*, and offsetting them by 7.5 millimeters so that we have a larger avoid zone to begin with:

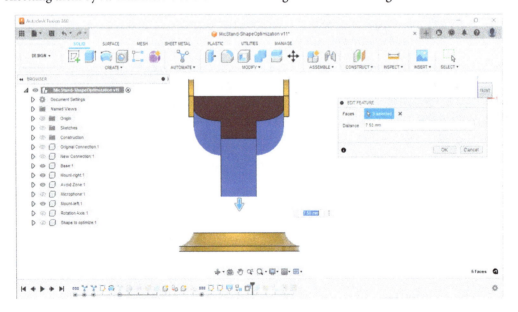

Figure 3.19 – Using the offset face command to create a larger body

Next, we can use the **COMBINE** feature within the **MODIFY** panel and remove the **Avoid Zone** body from the **Shape to optimize** body, as shown in *Figure 3.20*. This will ensure that we have enough space for the microphone and the two cables to rotate comfortably along the *X* axis.

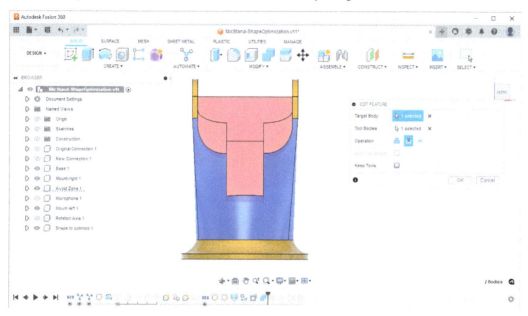

Figure 3.20 – Creation of the target body for shape optimization

Next, we will switch from the **DESIGN** workspace to the **SIMULATION** workspace and create a new shape optimization study. During the creation of the shape optimization study, we can also edit its settings and choose a smaller mesh size (**3%** in **Model-based Size**), as shown in *Figure 3.21*. Using a smaller percentage produces a finer mesh, which takes longer to solve but produces better results.

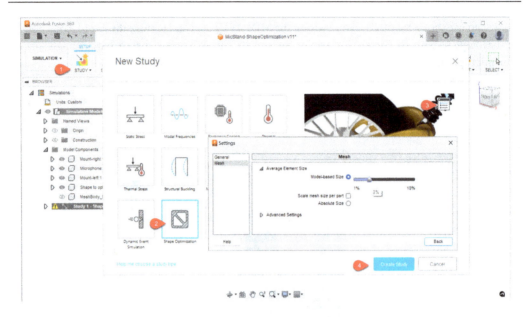

Figure 3.21 – Creating a mesh optimization study and customizing mesh settings

After creating a shape optimization study, we still need to edit our model further so that we are ready to solve the simulation. To accomplish this, we can enter the **SIMPLIFY** environment accessible by selecting the **SIMULATION** workspace, the **SETUP** tab, and the **SIMPLIFY** panel. Once we are in the **SIMPLIFY** environment, we can select all the bodies we do not need for the shape optimization study, right-click, and remove them, as shown in *Figure 3.22*:

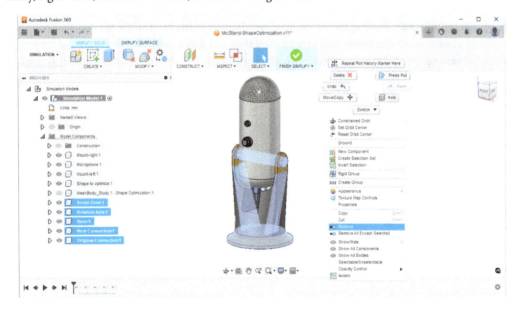

Figure 3.22 – Removing unnecessary bodies within the SIMPLIFY environment

Once we are left with all the bodies that will participate in the shape optimization study, it is time to define all the necessary inputs. We can start by designating the body named **Shape to optimize** as **Target Body**. Next, we can select the microphone and convert it to a point mass that will act on the top two surfaces of our target body. Next, we need to set the study materials. In this case, I have chosen **Aluminium AlSI10Mg** from within Fusion 360's Additive Material library.

Next, we need to assign the loading conditions. As shown in *Figure 3.23*, we need to define our loads in three distinct load cases. Load case 1 will reflect the scenario where the microphone is in the vertical position and is exerting a downward force on the stand. Load cases 2 and 3 represent the microphone being tilted forward and backward, respectively. In load cases 2 and 3, we also need to apply a moment to simulate the load the microphone would be exerting on the stand as its center of mass will be away from the mounting brackets along the Y axis. All three load cases also require a fixed boundary condition on the bottom of the base plate.

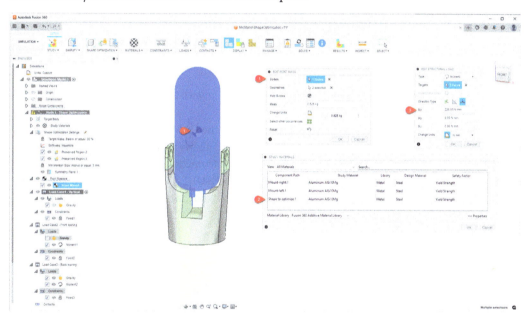

Figure 3.23 – Material definition and loading conditions for shape optimization of a microphone stand

Once our loads and boundary conditions are set, we can move on to designating certain bodies as preserved regions. In this case, we will utilize the bodies named **Mount-right** and **Mount-left** and create two box-shaped preserved regions based on them, as shown in *Figure 3.24*. We can also create a symmetry plane along the Global YZ plane and request that our final shape be symmetrical. In this shape optimization study, we are looking to reduce the mass of the starting shape by 70% and reach a target mass of 30% or less while maximizing the stiffness. We also defined the minimum member size as 3 millimeters or more, as a global constraint to our shape optimization criteria.

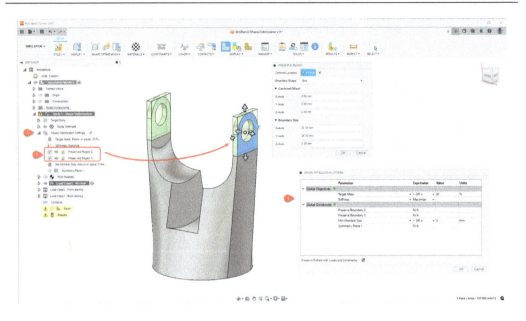

Figure 3.24 – Preserved regions and Shape Optimization criteria

After solving the shape optimization study, we will end up in the **RESULTS** environment within the **SIMULATION** workspace, as shown in *Figure 3.25*. The legend allows us to manipulate the optimized shape. Using a slider within the legend, we can change the outcome to be thinner or thicker, but the default outcome will be a target mass of 30%. Within the **RESULT TOOLS** panel, if we select the **PROMOTE** command, we will have the option to promote the outcome of the shape optimization as a new body to the **DESIGN** workspace.

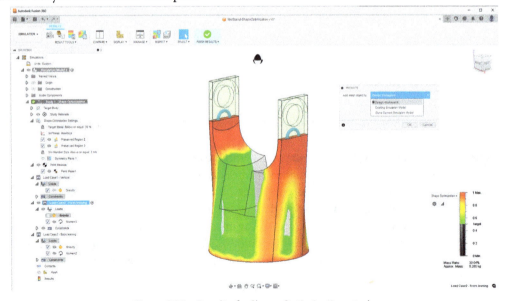

Figure 3.25 – Result of a Shape Optimization study

The body we promoted from the **SIMULATION** workspace will be available in the **DESIGN** workspace as a new mesh body. As shown in *Figure 3.26*, this new mesh body will have a unique name referencing the shape optimization study.

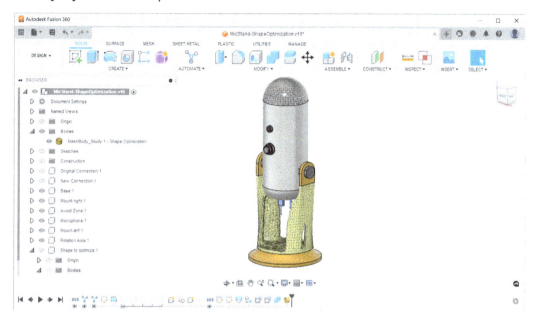

Figure 3.26 – Mesh outcome of a shape optimization promoted design workspace

Once a mesh outcome has been promoted to the **DESIGN** workspace, we have three options to use this new model. Our first option is to keep the mesh body as a mesh body and simply 3D print it as is. Our next option is to try to reverse engineer this mesh body using the various mesh section-based sketching and solid body creation functionality we covered in *Chapter 2*. Our third option is to try to convert the mesh body into a solid body.

If we want to pursue this third option, we will need access to the **Organic** mesh conversion method, as outcomes of shape optimization tend to be organic in nature. You can access the **CONVERT MESH** command within the **DESIGN** workspace by selecting the **MESH** tab followed by the **MODIFY** panel. The **Organic** method, as shown in *Figure 3.27*, is available to those who have access to the Design extension.

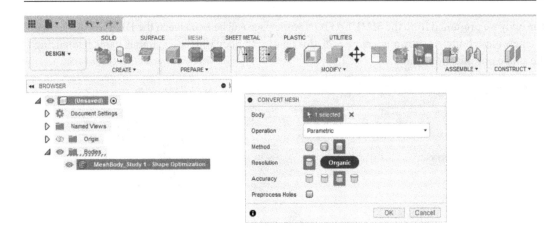

Figure 3.27 – CONVERT MESH command with the Organic method

Depending on the resolution and the accuracy we choose, we can convert this mesh body into a solid object using the **Organic** mesh conversion. *Figure 3.28* shows an example of what such an outcome would look like.

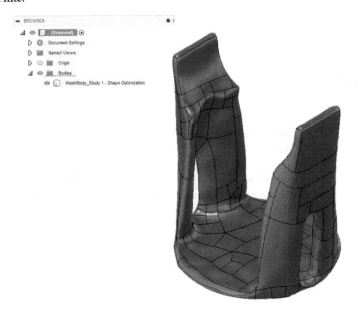

Figure 3.28 – The outcome of converting a mesh to a solid using the Organic method

To finish our design, we will need to utilize the **COMBINE** command one last time and unify multiple bodies representing the base, the mounting brackets, and the shape-optimized outcome of our target body, as shown in *Figure 3.29*.

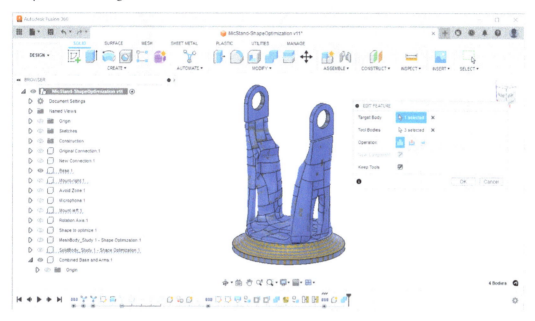

Figure 3.29 – Combining multiple bodies to make a single microphone stand

If we had chosen the first option, we could have left the shape optimization outcome as a mesh body and simply tessellated the base and the mounting brackets on the left- and right-hand sides. Then, we could have combined all four bodies as mesh objects and achieved a similar outcome. However, if we had utilized the mesh-based workflow and proceeded to create our final design as a mesh object, even though we would be able to 3D print our final design, we would not be able to simulate it using linear static stress within Fusion 360. This is because Fusion 360 requires objects to be solid bodies for them to be simulated. *Figure 3.30* shows our final design from four different angles. In this finished version, I have also moved all the unnecessary components into a new sub-assembly called **Extras** and turned their visibility off.

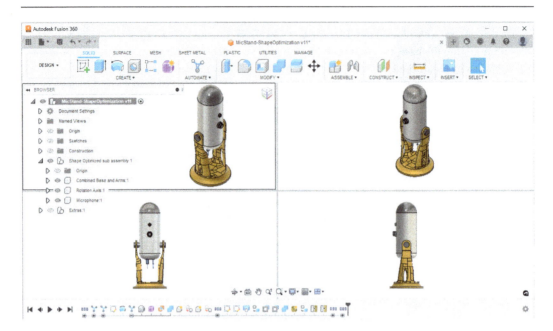

Figure 3.30 – The outcome of a shape-optimized microphone stand

In this section, we have introduced the concept of shape optimization and how it can be utilized for creating organic and lightweight geometry. We went through an example of how to create a target body for shape optimization and how to set up a simulation study. Upon conducting the study, we promoted our outcome to the **DESIGN** workspace and combined our outcome with existing bodies to create the final shape.

In the next section, we will conduct a linear static stress simulation to see whether the outcome of the automated modeling can withstand operating loads and whether we need to make any design adjustments to eliminate potential failures during daily use.

Structural simulation to detect and fix common failures

In the previous sections, we created several design alternatives using automated design and shape optimization. In this section, we will highlight how to simulate one of those design alternatives to check whether it can withstand the loads during operating conditions. We will also see whether the material we have chosen is appropriate based on the loading and try alternate materials. We will also try to detect and fix common failure points such as stress concentrations.

Let's get started by making a new design and saving it with a new name. Once we have a new Fusion 360 design document, using the **Insert Derive** command within the **SOLID** tab's **INSERT** panel, we can insert bodies and components from other Fusion 360 design documents, as shown in *Figure 3.31*. In this example, we will be inserting multiple bodies from one of the previous microphone stand designs we created using Automated design. Once we have four key bodies (**Base**, **Mount-left**, **Mount-right**, and **New connection**) inserted into this design, we can utilize the **COMBINE** command located in the **MODIFY** panel. We can also assign material for this design within the **MODIFY** panel's **PHYSICAL MATERIAL** function. We can choose the same additive **Aluminium AlSi10Mg** material we used in the previous sections for our first simulation study.

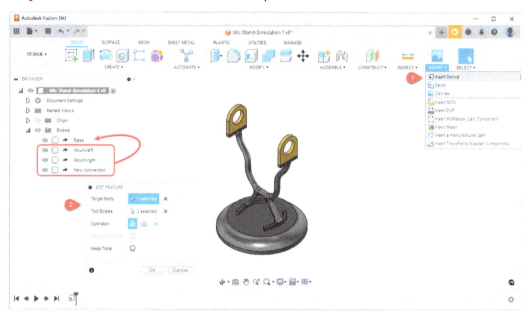

Figure 3.31 – Insert Derive and COMBINE commands to join multiple bodies to one

Now that we have the combined body within a single component, we are ready to start setting up our simulation. The first thing we will need to do is to switch from the **DESIGN** workspace to the **SIMULATION** workspace and create a new **Static Stress** study. Within this study, we will have to create three load cases, similar to how we set up our shape optimization study in the previous section. In each of these load cases, we will have to assign the same loads and boundary conditions we created for the shape optimization study.

The first load case will help us simulate the microphone standing in the vertical position. The loading we will choose for this load case will be two bearing loads acting on the cylindrical holes of the **Mount-left** and **Mount-right** sections of the design.

Load cases 2 and 3 will represent the microphone in a front-leaning and back-leaning position, just as how we had the shape optimization study set up in the previous section. For all three load cases, we will assign a fixed boundary condition at the bottom of the base. Next, we will assign material for the study. We will use the same material in the simulation study as the design material, as shown in *Figure 3.32*.

Figure 3.32 – Bearing load and study material for a linear static simulation

Once we have all our loads and boundary conditions assigned, we are ready to conduct this simulation study. We can start the solution from the **SETUP** tab's **Solve** panel.

> **Important note**
>
> In order to run a static stress simulation, you will need a trial, Commercial, Educational, or Startup license. Personal licenses do not have access to simulation capabilities: `https://www.autodesk.com/products/fusion-360/personal`.

Once the results of a static stress analysis become available, Fusion 360 will switch from the **SETUP** tab to the **RESULT** tab of the **SIMULATION** workspace. You will be able to see your simulation results and change to various result types. By default, Fusion 360 displays the factor of safety results. As shown in *Figure 3.33*, if you switch the result type to **Stress**, using the dropdown next to the legend, you can see the **Von Mises** stress this part will experience under each load case.

> **Important note**
>
> Von Mises stress is a universally accepted method for calculating an equivalent stress value from the six stress tensors (*XX, YY, ZZ, XY, YZ,* and *ZX*). It does not represent the true stress a given part experiences, but it is one method we can use to calculate the factor of safety for a given part and determine whether parts made from ductile materials, such as metals, will yield or fracture under certain loading conditions.

In this example, we will focus on load case 2, which represents when the microphone is in the front-leaning position. If you turn on the minimum and maximum probes using the **Show Min/Max** command located in the **INSPECT** panel, you can see where the highest stress will occur on this part.

It looks like we have high stress close to where an organic shape meets a prismatic shape close to the bottom of this part. Even though the stress results are low, and the factor of safety results are high, the location where the max stress is being displayed is a cause for concern and could be a potential failure point depending on which material we print this object with and/or the 3D printing method we choose.

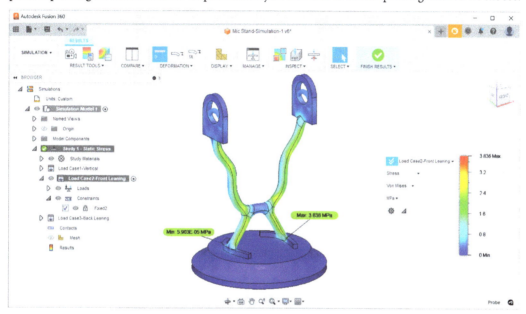

Figure 3.33 – Results of linear static stress for an initial simulation

To take this idea a little further, let's duplicate **Study 1** and create **Study 2**. In **Study 2 - Static stress**, we will choose a different material but keep all the loads and boundary conditions the same for all three load cases. The ability to change materials is one of the main benefits of conducting multiple studies using study duplication workflows within Fusion 360.

In **Study 1**, we chose aluminium as our material; in **Study 2**, let's change the study material to **PA 12**, which is a plastic material that can be printed using an HP MultiJet Fusion 3D printer. *Figure 3.34* shows the physical and thermal material properties of this material.

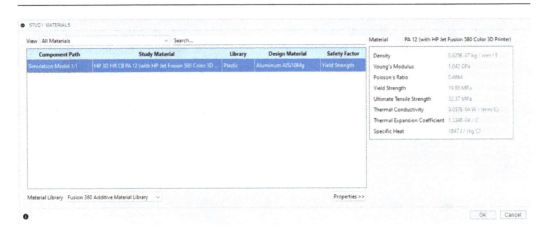

Figure 3.34 – Physical and thermal properties of PA 12 as printed on HP Jet Fusion 580

After running the second study, we can compare the results of **Study 1** to **Study 2**. Using Fusion 360's **COMPARE** functionality, we can split our layout into a multi-layout view, as shown in *Figure 3.35*.

In this figure, we can see two images on the left that are from **Study 1**, which used aluminium as the material. We can also see two images on the right that are from **Study 2**, which uses PA 12 as the material. The top two images show the factor of safety results. The bottom two images show the stress results for load case 2. Even though the stress results are relatively small for both simulations, the factor of safety drops from 8+ down to 5.3 when using a plastic material versus aluminium.

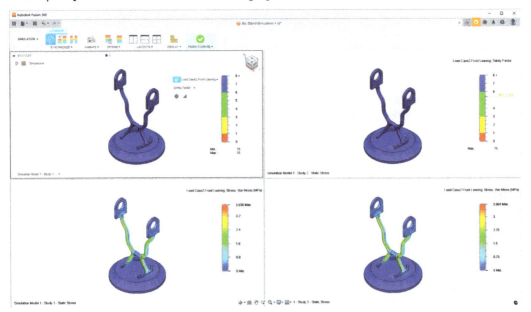

Figure 3.35 – Comparing the results of two simulations with different materials

Upon closer inspection, the high stress is in the same location as before. As mentioned earlier, this is an area where there is a chance of stress concentration. Stress concentration is a location within a model where the stress tends to rise regardless of the mesh size we use to simulate our model. It can be caused by a localized force, such as a point load, or an abrupt change in geometry. If there is a stress concentration in a model, it could lead to a fracture point during operation. If possible, it is always a good idea to either change the loading conditions or the geometry to remove stress concentration zones.

Let's see whether we can address the problem by making a design change. To make a design change, we will have to go back to the **DESIGN** workspace. In order to not lose the simulation results for the first two studies, let's save our design with a new name. In this case, I have chosen to name my design `Mic Stand-simulation 2`.

Figure 3.36 shows a zoomed-in version of the region where the organic legs meet the prismatic feature. One way to eliminate stress concentration is to add fillets. If we try to add fillets to the geometry in its current condition, Fusion 360 fails to create the fillet on our selected edges due to the complexity of the current shape.

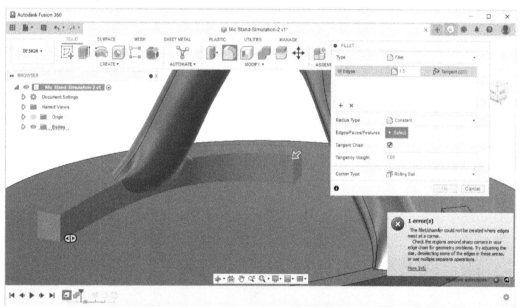

Figure 3.36 – Error during fillet creation for the problem region

Such errors during fillet or chamfer creation can occur in CAD software that utilizes **boundary representation (BREP)** modeling. Whenever we run into such errors, we may have to come up with a different solution to change our design, so that we can resolve the stress concentration issue and create smooth transitions between edges. In this case, I have chosen to construct a new prismatic body. We can do that by creating a new sketch on the top of the base and using an offset from the existing prismatic shape of 5 millimeters, as shown in *Figure 3.37*.

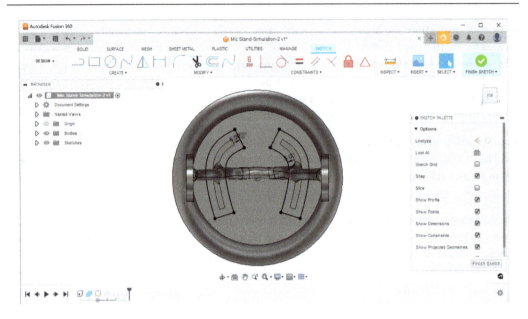

Figure 3.37 – Sketching two new profiles with offsets from the prismatic shapes

We can extrude two new shapes by selecting these new profiles. This will be the new prismatic base that connects to our organic stand. In order to save weight, we can create our extrude with a taper angle. In this example, we will be utilizing a 20-degree taper angle as a part of the extrude feature, as shown in *Figure 3.38*.

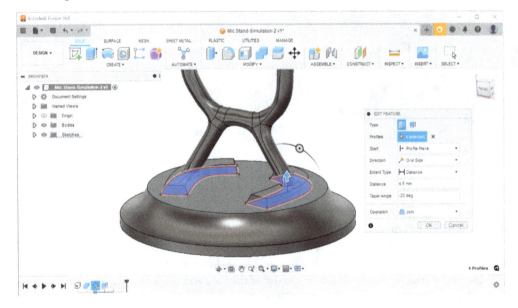

Figure 3.38 – Extruding two new prismatic shapes with a taper angle

Once we have our new prismatic base, we can choose all the faces and/or edges where the base meets the organic shape and create a new transition using the **FILLET** command located in the **MODIFY** panel. In this example, we'll be adding a 1.5-millimeter radius fillet, as shown in *Figure 3.39*:

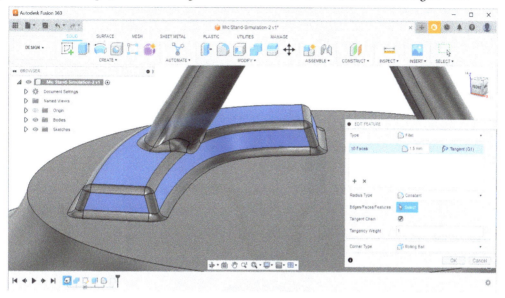

Figure 3.39 – Applying a fillet feature to the prismatic base and the organic stand

While we're focusing on adding fillets, it is also a good idea to add two fillets to the sharp corners at the bottom of the left and right mounting zones. As shown in *Figure 3.40*, we can reduce weight and make our design safer to use if we go with a larger fillet radius.

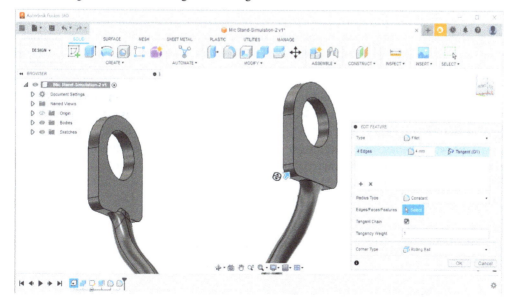

Figure 3.40 – Applying a 4 mm fillet to the mounting brackets

Now that we have made all of the necessary changes to our model, we can switch to the **SIMULATION** workspace. After we make sure that we have the correct PA 12 material chosen, we can rerun the static stress study. This new study solved within this new Fusion 360 design shows a much more balanced stress distribution, as can be seen in *Figure 3.41*. It no longer has its highest stress point where the organic stand meets the prismatic base. It also has its highest stress value reduced compared to the previous designs/simulations.

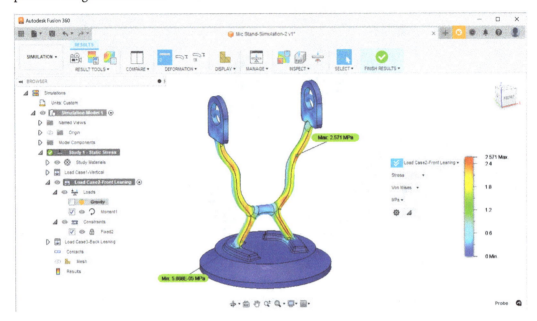

Figure 3.41 – Stress results after a design change to eliminate stress concentration

Having gone through these three simulations, we can be confident that not only will our microphone stand survive daily use but, when 3D printed, it will not have potential weak spots, which could lead to fracture over time.

It is important to note that we have simulated these parts using materials with an isotropic material model assumption. The materials in Fusion 360's Additive Material library, such as metals, powders, and resins, have physical properties based on having been 3D printed and tested. In the case of materials such as PA 12, which behave in an orthotropic fashion, Fusion 360 uses build direction mechanical properties and follows an isotropic material behavior assumption. For additional details on each material and 3D printing technology, please refer to the spec sheets of the material being simulated – for example, `https://h20195.www2.hp.com/v2/getpdf.aspx/4AA7-1533ENA.pdf`.

In conclusion, there are a number of ways we can lightweight parts. Fusion 360 offers several easy-to-use methods for lightweighting existing parts as well as creating new designs.

Summary

In this chapter, we have created organic and lightweight models using Automated Modeling. Automated Modeling gave us six design alternatives to choose from and, once we chose one of them, we were able to modify its volume and create an organic and lightweight design alternative. Next, we designed a starting shape and used Shape Optimization to lightweight that target shape based on loads and boundary conditions.

However, neither of those methods created shapes based on stress results. In order to make sure that our chosen design alternative could withstand stresses that occur during operating loading conditions, we conducted a static stress study. Our initial simulations pointed out that our material choice was way too strong for our needs. After changing our material from a metal to a plastic, we realized that our design would still perform well for its intended use. However, we also identified a stress concentration zone and conducted one final design change to eliminate the stress concentration in order to minimize potential future failure points.

Using a similar workflow, you can be sure to create lightweight and compliant designs that not only can withstand operating conditions but will have a long life in the field. In the next chapter, we will focus on how to further lightweight designs using various hollowing and latticing methods to reduce material and energy consumption.

4

Hollowing and Latticing Parts to Reduce Material and Energy Usage

Welcome to *Chapter 4*. As we mentioned in the previous chapter, one key benefit of Additive Manufacturing is the ability to create lightweight parts. In this chapter, we will cover several tools we have within Fusion 360 for hollowing bulky parts and applying lattice structures to the hollowed-out sections to create lightweight versions that we can 3D print, using reduced material and energy.

We will start by talking about the various 3D printing technologies and how we can hollow parts depending on the technology. After that, we will focus on creating drainage holes so that the parts we 3D print using resin or powder can be removed once the printing process is complete. Finally, we will discuss the various tools we have at our disposal for creating lattices.

In this chapter, we will cover the following topics:

- Hollowing parts
- Creating drainage holes
- Creating solid and volumetric lattices

By the end of this chapter, you will have learned how to hollow solid and mesh bodies. You will be able to model drain holes, and you will have a good understanding of how to create a lattice shape to fill the inside of the hollowed-out components.

Technical requirements

All the topics covered in this chapter are accessible to the startup and educational Fusion 360 license types. If you have a commercial license of Fusion 360, in order to create and modify a volumetric lattice, you will also need access to the Design extension (`https://www.autodesk.com/products/fusion-360/design-extension`)

Autodesk offers a 14-day trial for the Design extension. You can activate your trial within the extensions dialog of Fusion 360.

Design extension access is not available for personal/hobby licenses of Fusion 360.

The lesson files for this chapter can be found here: `https://github.com/PacktPublishing/3D-Printing-with-Fusion-360`.

Hollowing parts

In previous chapters, we covered how to import and edit existing designs made up of solid and mesh bodies. We also utilized shape optimization and automated modeling to have Fusion 360 generate lightweight design alternatives. Then, we simulated those geometries to see whether the generated models can withstand the operating loads. During all these activities, our main objective was to create models that can be 3D printed.

We can use 3D printing to manufacture end-use parts, as well as quickly create prototypes before mass production using other manufacturing methods. Regardless of why we choose 3D printing as our method of manufacturing, we all want to produce our physical models quickly and efficiently.

As outlined in *Chapter 2*, there are multiple 3D printing technologies available. Each 3D printing technology has unique design requirements to produce physical models efficiently. When you are printing parts that do not require structural strength, hollowing your model can be a great way to save time and material.

When manufacturing large-volume components using **fused filament fabrication** (**FFF**) 3D printers, it is common practice to utilize multiple contour lines to make the part strong. In FFF, it is also common to utilize an infill pattern with a density that could range from 15 to 30%, to keep the parts lightweight. This same slicing technique does not directly work for other 3D printing technologies, such as **stereolithography** (**SLA**), Multi Jet Fusion, or **selective laser sintering** (**SLS**). When creating parts to prototype using those technologies, designers must proactively think about how to lightweight parts to reduce material and energy usage.

In this section, we will focus on how to hollow parts for 3D printing technologies such as stereolithography, Multi Jet Fusion, and selective laser sintering. In most cases, hollowing out parts will save a significant amount of material instead of printing the entire part as a 100% solid object. Hollowing a part can also improve our printing success in resin-based printers (stereolithography and **Digital Light Processing** (**DLP**)) as it reduces the weight of the parts being printed and decreases the stresses on the build plate so that parts do not detach from the build plate during the printing process.

While hollowing parts, we also have to make sure our wall thickness is large enough for our chosen printing process so that our parts can be manufactured successfully and do not break during their intended use.

When working with solid models, Fusion 360 offers two methods for hollowing parts. Both methods can be accessed with the same **SHELL** command. The **Shell** command is located within the **DESIGN** workspace | the **SOLID** tab | the **MODIFY** panel, as can be seen in *Figure 4.1*.

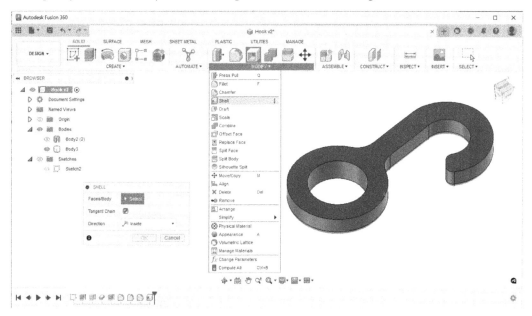

Figure 4.1 – Accessing the shell command within the SOLID tab

Fusion 360's **SHELL** functionality is very versatile. It allows you to select faces or bodies as input. If you select a face and then specify the thickness, the selected face will be removed, and you will end up with a hollow version of that body, as can be seen in *Figure 4.2*. The **SHELL** dialog also allows you to select multiple faces. By pressing down the *Shift* key, you can select multiple faces, and the shell command will remove those faces similarly.

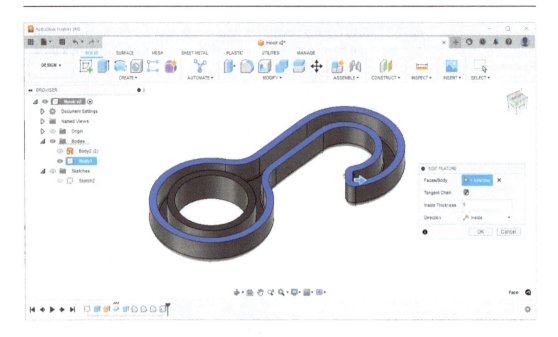

Figure 4.2 – Selecting a single face to turn a solid body into a hollowed version

Another way to use the **SHELL** command is to select a body instead of a face(s). After accessing the **SHELL** dialog, you can simply select the body to shell from the browser. Alternatively, you can select the body to shell from the canvas. If you want to utilize the canvas, you have several options to select a body. One option is to change **Selection Priority** to **Select Body Priority**. After making that change, you can simply point and select the body you wish to shell, as shown in *Figure 4.3*. Another option is to access the **Selection Filters** pull-out menu and make sure Bodies is the only item that is checked, which will allow limiting your Shell selection method to a Body.

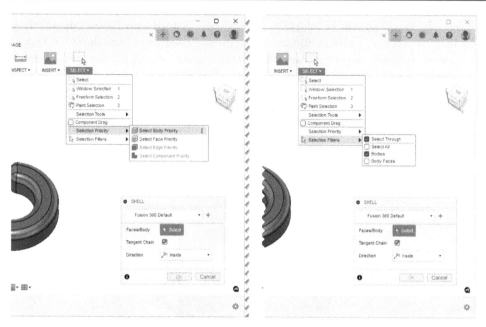

Figure 4.3 – Using the canvas to select a body to shell

Once you execute the shell command on the body, you will typically not see any visual differences in how your model looks. There are several ways you may want to inspect your model to get a better understanding of what the shelled outcome looks like. The first method is to simply change your visual style to **Shaded with Hidden Edges** or **Wire Frame with Hidden Edges**, as shown in *Figure 4.4*. These visual styles will both give you an indication that the model has been shelled by displaying additional edges on the inside of the model.

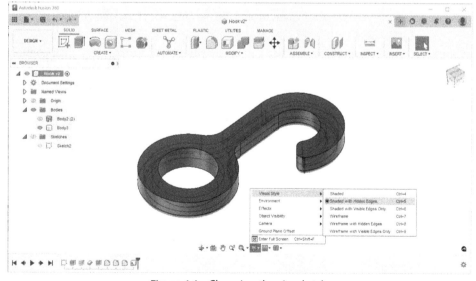

Figure 4.4 – Changing the visual style

If you do not want to change your visual style from the default **Shaded with Visible Edges Only**, you may want to utilize **Section Analysis**. You can access the section analysis from the **SOLID** tab | the **INSPECT** panel. After selecting a cut plane, the section analysis command allows you to translate that plane along its normal direction to show you a slice view of the component. You can also flip the cut direction so that you can look through the part as needed. *Figure 4.5* shows a **Section Analysis** view for this part. You will also notice that the section analysis adds a line item called **Section1** to the browser within the **Analysis** folder. We can control the visibility of all section analyses, as well as other analyses we have added to our model, by toggling the visibility icon.

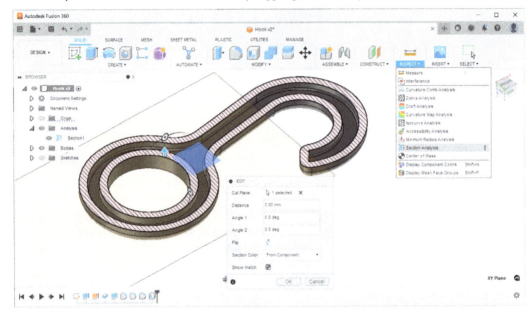

Figure 4.5 – Applying a section analysis to visualize the outcome of the shell command

If we are working with a mesh object, we also can create a hollowed version. The shell command within the **MESH** tab | the **MODIFY** panel allows us to select the body, enter a thickness, and create a hollowed version of that object. Unlike the **SHELL** command, which is available within the **SOLID** tab, the **SHELL** command within the **MESH** tab does not allow the selection of faces. This means that we cannot create the same shapes using the **SHELL** command within the **MESH** tab. However, in situations where the shell command within the **SOLID** tab fails to create the desired outcome, you can tessellate your solid body and create a mesh version of it. After that, you can try the shell command within the **MESH** tab, as shown in *Figure 4.6*.

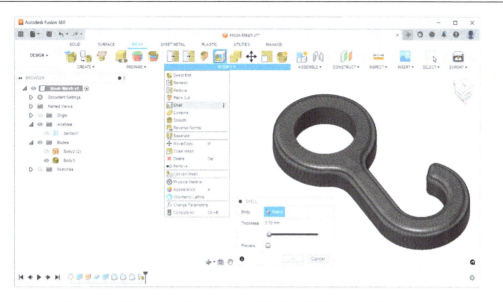

Figure 4.6 – Using the shell command within the MESH tab to hollow a body

Once we have the hollowed version of the mesh body, we can review the outcome by changing our visual style to shaded with hidden edges or by applying a section analysis similar to how we added a section analysis to a solid body in *Figure 4.5*. When we add a section analysis to a mesh body, we will not see the hatch view. Instead, we will see the triangles of the mesh body sectioned along the cut plane, as shown in *Figure 4.7*.

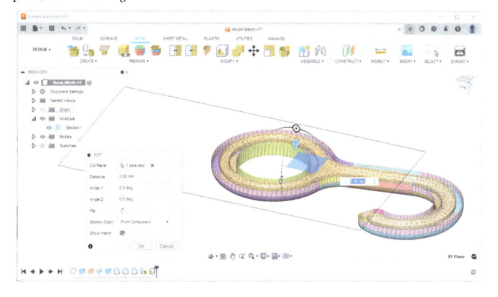

Figure 4.7 – Section analysis view of hollowed-out mesh body

In this section, we focused on how to use the shell command on both solid and mesh bodies. We covered how to shell a solid body either by removing one or more of its faces, as well as keeping the body intact and simply hollowing out the inside volume of the model. Using a mesh-specific shell command, we repeated the process on a mesh body. After each shell operation, we learned how to visualize the resulting shelled body by changing the visualization settings, as well as adding a section analysis to see the interior surfaces and hatching on the cut plane where appropriate.

In the next section, we will go over how and where to add drain holes to hollowed-out models in order to remove any raw material that would be trapped inside during and after 3D printing such models.

Creating drainage holes

After 3D printing hollowed-out parts, we will need to drain or remove all trapped powder/resin during the postprocessing phase. That's why we need to include drainage holes in our design phase. The diameter of the drainage holes depends on the 3D printing technology we are utilizing as well as the viscosity of the material we are 3D printing with. For example, if you are using a laser-based resin printer (SLA), your resins are generally more viscous, compared to resins for DLP printer resins. Certain resins can also be used for both DLP and SLA printers. The more viscous a resin is, the larger the diameter of the drainage hole needs to be.

Table 4.1 shows two resins from different sources with varying viscosity. According to the manufacturer, the second resin is compatible with both SLA and DLP printers:

Resin	Viscosity @ 25 deg. C
Prusa Orange Tough: `https://www.prusa3d.com/product/prusament-resin-tough-prusa-orange-1kg-2/#downloads`	180-280 centipoise
Liqcreate Strong-X: `https://www.liqcreate.com/product/strong-x/#technical`	550 centipoise

Table 4.1 – Resins and their viscosities for SLA and DLP printing

The smallest-hole diameter the technology can create using SLA 3D printing is 0.5 millimeters, as stated in the design rules for 3D printing on Hubs.com, as shown in *Figure 2.25* of *Chapter 2*. The same chart also shows that the minimum diameter of a hole for a SLS process is 1.5 millimeters. However, it is common practice to introduce a minimum of two 2-millimeter diameter drainage holes for DLP printers, and three or more 3.5-millimeter diameter drainage holes for SLA printers. Similarly, for SLS printers, a minimum of two 3.5-millimeter drain holes is recommended for the best results. Remember that adding more and larger drain holes always makes it easier to remove un-sintered powder as well as uncured resin from internal cavities.

The placement of drain holes also requires consideration. Another reason we need drain holes, especially for resin-based printers, is to eliminate the cupping effect. The cupping effect occurs when we 3D print a layer and the print bed moves up and down within the resin, building up pressure within the enclosed volumes. Adding a drain hole to relieve pressure where cupping may occur will help our prints be successful, eliminate support structure failures, and remove layer-shift problems.

Now that we have covered why we need drainage holes and how big they should be, let's discuss where we generally should place them. In resin printing, it is best to place drainage holes as close to the build plate of the printer as possible in the orientation chosen to be for the part to be 3D printed. This means we have to orient our model for 3D printing before designing drain holes. We will cover arranging an orienting component in detail in *Chapter 8*. Once we have our orientation selected, we can place our drain holes on the bottom surface of our model close to the build plate. This will ensure that we don't have pressure building up in our hollow model during the printing process for the resin printers. We don't have the same issues about drain hole placement for SLS printers, as cupping and pressure buildup during printing is generally not an issue. This means we can place them anywhere on our model as long as they are away from critical surfaces.

Now let's demonstrate how we can create drain holes using Fusion 360. *Figure 4.8* shows the same hook model we worked on in the previous section already oriented to be printed using an SLA/DLP printer. To better visualize the orientation of the part with respect to the build plate, I have also included a construction plane along the XY plane. In addition, the part opacity is down to 50% to show that the part has been hollowed out.

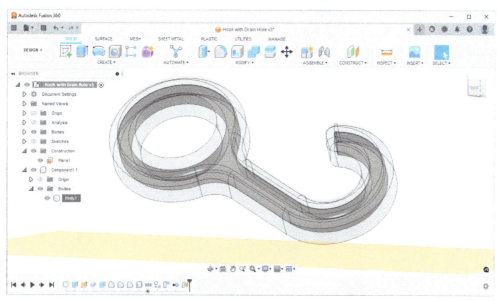

Figure 4.8 – Hollowed-out version of the hook model oriented for resin 3D printing

For this part, we will need a minimum of two 2-millimeter diameter holes. The best placement for these holes is at the base of the hook on the back face of the component, as shown in *Figure 4.9*. As this model allows for the creation of bigger holes, I have chosen the hole diameter to be 3 mm for both holes.

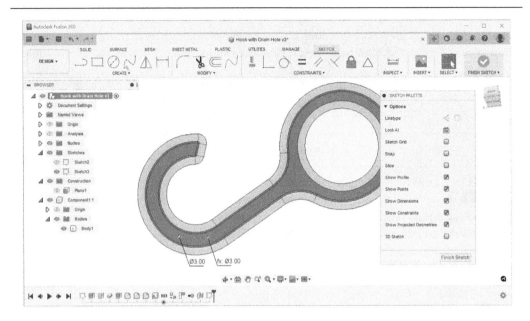

Figure 4.9 – Creating a sketch and drawing two circles

Once our sketch is done, we can create the drain holes using the **EXTRUDE** command or the **HOLE** command. In *Figure 4.10*, you can see how we can create the drain holes using the **HOLE** command and have detailed control over the size, depth, and shape of the hole.

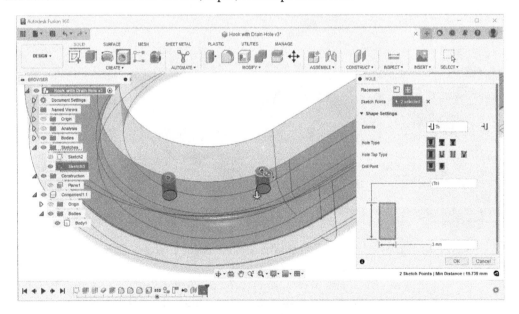

Figure 4.10 – Creating two drain holes using the HOLE command

After creating our holes, we can add a **SECTION ANALYSIS** as shown in *Figure 4.11*. During the creation of the section analysis, if we choose the XY plane from the Origin node within the **BROWSER** as the cut plane, offset it in the positive Z direction by approximately 22 millimeters, and flip the section view, we can see the bottom of the part and observe that we have created two small holes where the resin can drain during 3D printing:

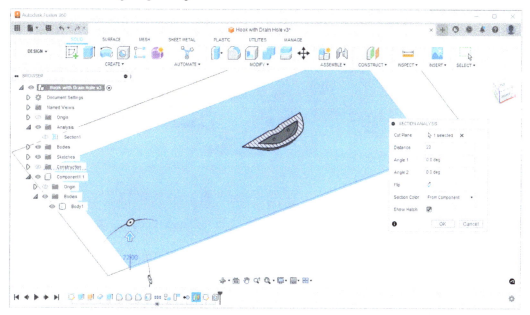

Figure 4.11 – Section analysis shows two drain holes

Creating drain holes for a mesh-based model is similar to creating them for a solid model. After opening our mesh body, hollowing it out, and rotating it to the orientation we wish to 3D print it in, we will have to access the **DIRECT EDIT** command located within the **MESH** tab | the **MODIFY** panel, as shown in *Figure 4.12*:

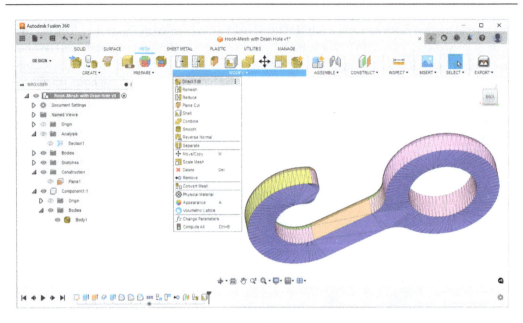

Figure 4.12 – Utilizing the Direct Edit mode for a mesh body

Once we are in the **DIRECT MESH EDITING** tab, we can create a plane through three points using the functionality located within the **CONSTRUCT** panel. We will later utilize this construction plane to create a sketch so we know the location of our holes.

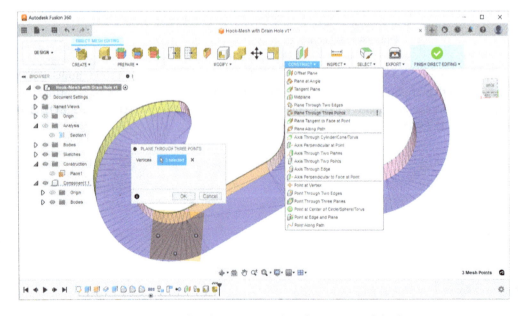

Figure 4.13 – Creating a construction plane on a mesh body

Once the construction plane has been created, we can select the **FINISH DIRECT EDITING** command in the ribbon, which will get us back to the parametric design mode of Fusion 360. Now we are ready to create a sketch. We can access the **SKETCH** command within the **SOLID** tab | the **CREATE** panel. Once we are in the **SKETCH** tab, we will need to create multiple circles to represent the drain holes. *Figure 4.14* demonstrates two circles 20 millimeters apart from each other that have 3-millimeter diameters. Once our sketch is done, we can select the **FINISH SKETCH** command in the ribbon:

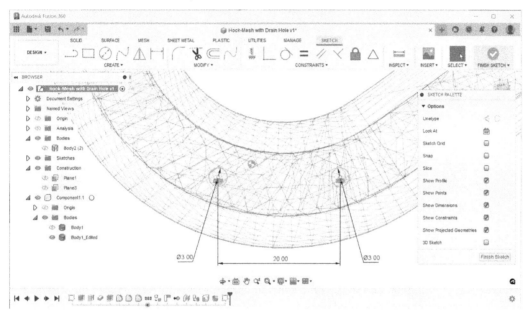

Figure 4.14 – Creating two circles within the sketch environment

Now, we are ready to extrude those two profiles and create two cylinders. Using the **EXTRUDE** command, located within the **SOLID** tab | the **CREATE** panel, we can select our newly created circular profiles and extrude them so that they are slightly taller than the thickness of the part. In the previous section, when we were hollowing out this part, we designed the wall thickness to be 5 millimeters. Therefore, creating a 10-millimeter extrude should be sufficient for our holes.

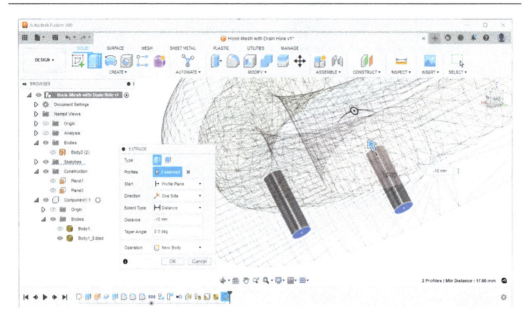

Figure 4.15 – Extruding circular profiles to make cylinders to represent drain holes

In *Chapter 2*, we talked extensively about how to work with CAD and mesh files using Fusion 360. One of the workflows we covered in that chapter was tessellating solid models to convert the solid bodies into mesh bodies in order to combine them with other mesh bodies. That is exactly what we will do next. Let's start by using the **TESSELLATE** command located within the **MESH** tab | the **CREATE** panel, on the two new cylindrical solid bodies.

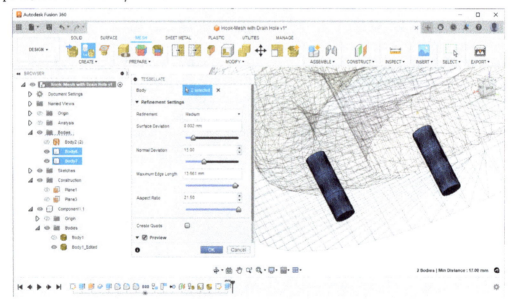

Figure 4.16 – Tessellating the two cylindrical solid bodies

Once we have converted the solid bodies into mesh bodies, using the **COMBINE** feature located in the **MESH** tab | the **MODIFY** panel, we can select the **REMOVE** operation and remove them from the direct edited version of the tessellated hook body, as shown in *Figure 4.17*:

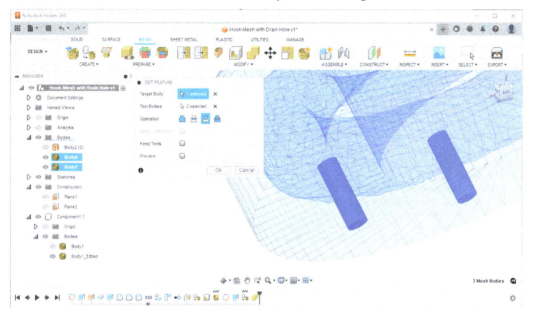

Figure 4.17 – Using the Combine command to remove the two cylinders from the hook

Just as we added a section analysis to the solid body to display the drain holes, we can apply a section analysis along the XY plane to the meshed body and move it up in positive **Z** by 22 millimeters to show the drain holes:

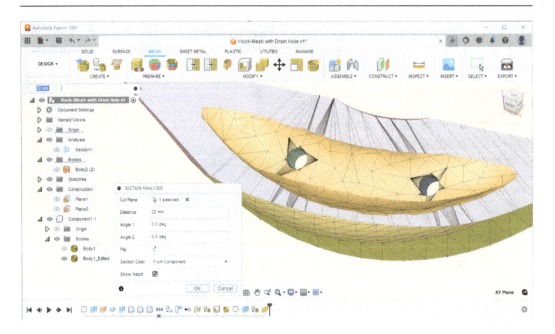

Figure 4.18 – Section Analysis to show drain holes within the mesh body

In this section, we covered how to work with solid bodies and mesh bodies to add drain holes once they are properly oriented for resin printing. The same principles also apply to adding drain holes to geometries that utilize other 3D printing technologies, such as SLS or Multi Jet Fusion.

Creating internal lattice structures

In previous sections, we covered various ways we can hollow solid or mesh bodies. We demonstrated how to add drain holes so that any trapped powder or resin can be removed after the part has been 3D printed. Our main goal in hollowing parts was to lightweight them so that we don't waste material and energy while printing our parts.

Whenever we create hollow objects, we should also pay attention to whether those parts are self-supporting to be 3D printed, based on the 3D printing technology we have chosen. You will remember from previous chapters that if we are 3D printing using SLS, we don't need support structures. However, if you are 3D printing using a resin-based printer, such as SLA or DLP, portions of your model with overhang angles above a critical angle will require support structures. We often think about support structure as material printed on the outside of a part so that it will support our part during printing. A similar concept also applies to parts that we hollow. If we hollow a part, we are essentially creating internal surfaces, which may also require support structures, so that our parts can be printed successfully with a resin printer. In this section, we will talk about how to create those internal support structures.

One method to create an internal support structure after hollowing a part is to create a lattice. Fusion 360 offers several different workflows for creating internal lattices. In this section, we will specifically talk about how to create volumetric, mesh-based, and solid lattices. In *Chapter 10*, we will cover how to create support structures within the context of the Manufacture workspace in more detail. In the same chapter, we will apply those support structures to the exterior surfaces of the parts, as well as interior surfaces as needed. After all, applying a support structure to the interior surfaces of a part serves the same purpose as adding an internal lattice.

Let's start our exploration of lattice generation using volumetric lattices. Fusion 360 allows users to create a volumetric lattice after accessing the Design extension. We have covered how to use the Design extension in *Chapter 2* and *Chapter 3* in the context of converting organic mesh bodies into solid objects. In this chapter, we will utilize another key functionality of the Design extension, **Volumetric Lattice**. **Volumetric Lattice** is available within the **MODIFY** panel of the **SOLID**, **MESH**, and **PLASTIC** tabs, as shown in *Figure 4.19*.

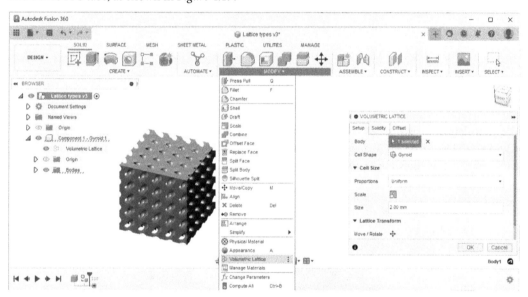

Figure 4.19 – A uniform gyroid lattice applied to a solid body

The **Volumetric Lattice** command changes a solid or a mesh body to create a volumetric representation of the model. Using this command, you can quickly add an internal lattice structure within the boundary of the body. *Figure 4.19* shows the application of a gyroid lattice with a cell size of 2 millimeters applied to a box with an edge length of 10 millimeters. In some cases, we may want to control the size of a given cell to stretch out the lattice in a certain direction. The **Volumetric Lattice** command has the option to change the proportions of the cell size from uniform to non-uniform. *Figure 4.20* shows a non-uniform cell size where the X and Y directions of the lattice cell are 3 millimeters, but the lattice cell is stretched on the Z direction and is set to 5 millimeters:

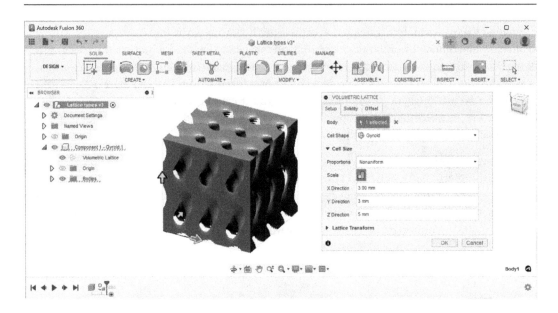

Figure 4.20 – Lattice cell is stretched in the Z direction with a non-uniform cell size

Depending on the cell shape we choose and our part's orientation, we may need to consider orienting and/or translating the lattice cell so it can be printed without needing support itself. For such cases, the **VOLUMETRIC LATTICE** dialog allows us to transform the lattice by positioning and orienting the reference cell, as shown in *Figure 4.21*:

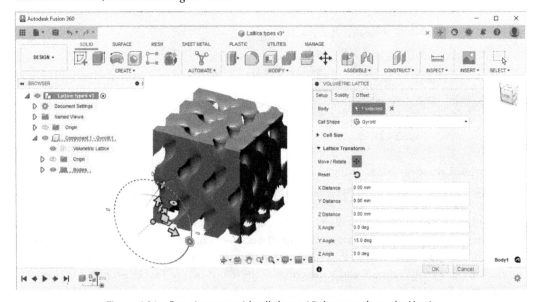

Figure 4.21 – Rotating a gyroid cell shape 15 degrees along the Y axis

The **VOLUMETRIC LATTICE** dialog is made up of three tabs. The **Setup** tab allows us to select a body and apply a cell shape to turn our solid or mesh body into a volumetric model. After selecting a body, it automatically applies a gyroid cell shape as that is the default lattice type. However, we can choose a different cell shape by simply expanding the dropdown and toggling to a different cell shape, such as the **Cross** shape shown in *Figure 4.22*.

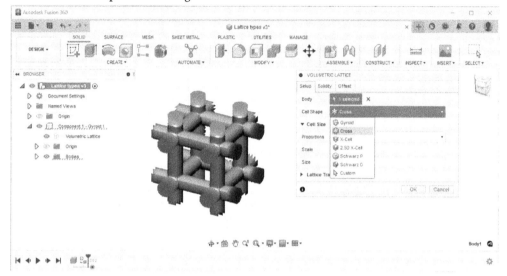

Figure 4.22 – Changing the cell shape within the VOLUMETRIC LATTICE dialog

The **VOLUMETRIC LATTICE** dialog within Fusion 360 has six default cell shapes. The outcome of all of these shapes applied to 10 by 10 by 10 boxes with a cell size of 10 and a solidity of 0.4 can be seen in *Figure 4.23*:

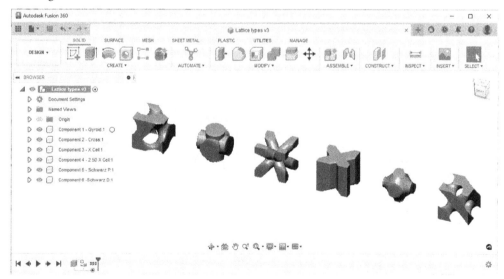

Figure 4.23 – The resulting shape of each cell shape within volumetric lattice

As the gyroid shape is a common shape in 3D printing, we will continue using it for the rest of this section. When we switch to the **Solidity** tab, the **VOLUMETRIC LATTICE** dialog offers us the option to modify the solidity of the lattice being applied to our reference body. The solidity input ranges from 0 to 1. A solidity of 0 means we will not have any material, and a solidity of 1 refers to a fully dense material distribution:

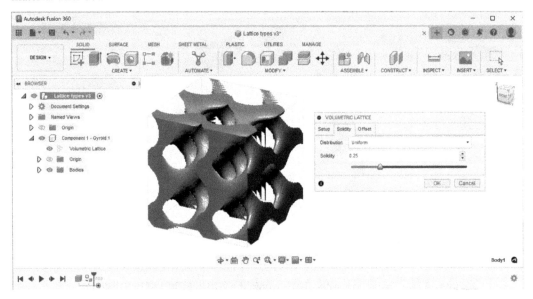

Figure 4.24 – Modifying the solidity uniformly for a volumetric lattice

We also can control how the solidity changes along a path. After changing the distribution type to **Gradient Along Path**, we can select an edge or a sketch line. The **VOLUMETRIC LATTICE** dialog will automatically detect the start and end points of the selected object and give us options to control the solidity between the start and end positions, as shown in *Figure 4.25*.

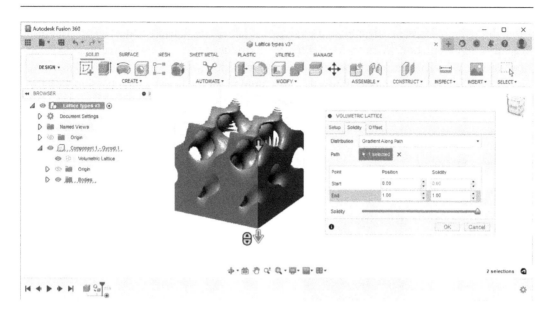

Figure 4.25 – Varying the solidity of a lattice along the edge of the bounding body

The final tab of the **VOLUMETRIC LATTICE** dialog is called **Offset**. Within this tab, we can select one or more geometry inputs, such as a face or an edge, and assign thickness and solidity. *Figure 4.26* shows the outcome after selecting the front-looking face of the cube and assigning a thickness of 2 millimeters with a solidity of 1. The resulting volumetric model represents a 2-millimeter-thick skin on the front-looking face, followed by a gyroid lattice.

Figure 4.26 – Assigning an offset face with a thickness

Up until this point, we have mostly utilized the volumetric lattice functionality to convert solid bodies into a lattice representation. With the **Offset** tab, we introduced the concept of creating a skin. If desired, we can go back to the **Solidity** tab and enter a solidity of 0, which in turn removes all material from the lattice portion of our model. If we take this action, all we will be left with is the skin we have applied using the **Offset** tab, as shown in *Figure 4.27*:

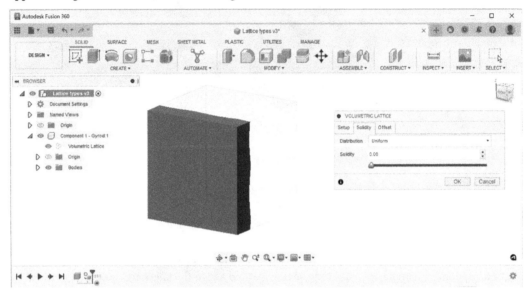

Figure 4.27 – Outcome of a 0-solidity lattice with a single offset face

We can take this one step further and apply the offset to all six faces of this solid cube. As shown in *Figure 4.28*, such an action will create a hollowed-out box with a thickness of 2 millimeters. Whenever we hollow a body, it is difficult to visualize as the exterior faces block us from seeing inside the object. Just as how we utilized the section analysis in the previous subsection, we can simply go to the **INSPECT** panel and add a section analysis along the XZ plane and offset it by a certain distance to take a look at what the inside of this volumetric body looks like.

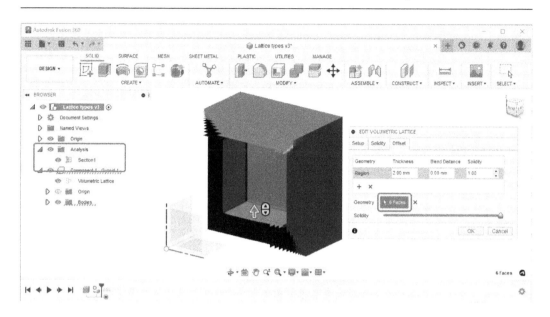

Figure 4.28 – Section analysis view of a hollowed-out volumetric body

As we covered in the previous subsection, whenever we hollow out a body, it is important to add drain holes so that the trapped resin or powder can be removed after 3D printing. If you are utilizing a volumetric lattice to hollow out a body, you can simply create a sketch, make a circle, and use the **HOLE** command to create a drain hole. In *Figure 4.29*, you can see the addition of a 2-millimeter-diameter drain hole.

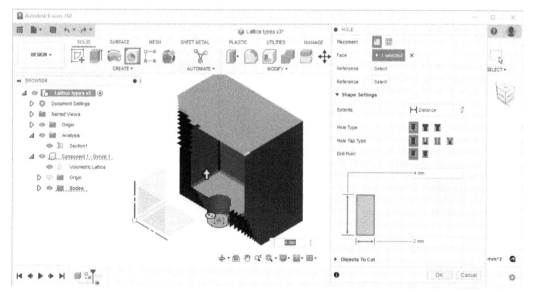

Figure 4.29 – Section view of a hollowed-out volumetric body with a drain hole

As the main objective of this subsection is applying lattices, let's go back to the **Volumetric Lattice** within the browser, right-click, and edit it, as shown in *Figure 4.30*. When we are editing our volumetric lattice, we can access the **Solidity** tab and increase the solidity from 0 to 0 . 2, thereby applying both a skin and a lattice to this part, which already has a drain hole:

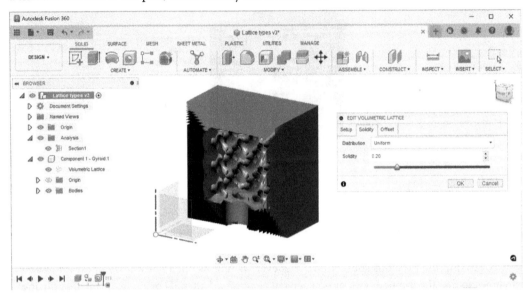

Figure 4.30 – Volumetric model shelled and latticed

Once we have our design finalized, Fusion 360 offers two methods for 3D printing a volumetric model. The first method is to switch to the manufacturing workspace, creating an additive setup with the appropriate 3D printer, and slice the volumetric model natively in Fusion 360 to be 3D printed. The second method is to create a mesh body so that we can export it to third-party slicers. To convert a volumetric body into a mesh body, we have to select it from the browser and right-click. As shown in *Figure 4.31*, after choosing the **Create Mesh** option, we will be able to create a mesh representation of the volumetric body:

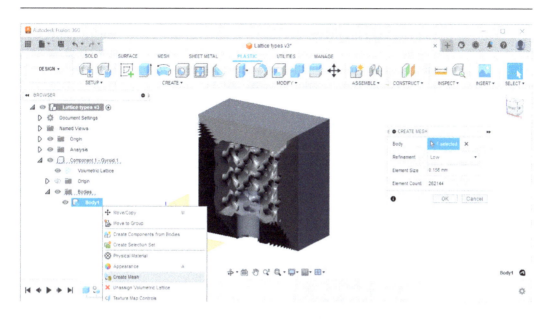

Figure 4.31 – Creating a new mesh body out of a volumetric body

The resulting mesh can be edited and visualized using all the native mesh tools we covered in *Chapter 2*. In *Figure 4.32*, we can see the resulting mesh body with a section analysis applied to show the internal lattice and drain hole.

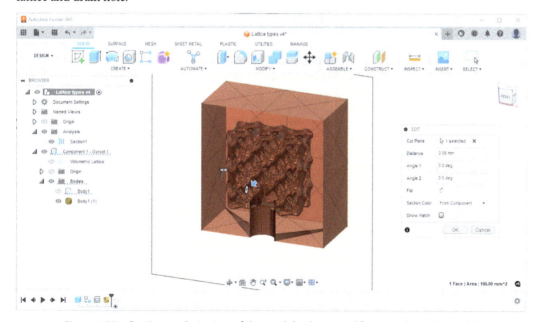

Figure 4.32 – Section analysis view of the mesh body created from a volumetric model

Now it's time to switch gears from the default cell shapes available within the volumetric lattice to custom shapes and solid lattices. As Fusion 360 is a parametric CAD tool, we can quickly generate a sketch, surface, or solid body, which we can then utilize as input for a custom cell type within the volumetric lattice. *Figure 4.33* shows the recreation of a cross cell shape using a sketched circle extruded along a distance and patterned twice, using a circular pattern to create three bodies, which were then combined to create a single-body lattice cell.

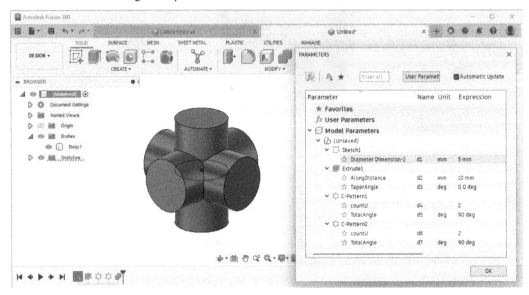

Figure 4.33 – Solid modeling of a single cell

Now that our custom cell has been designed, we can apply this cell shape to a different body within the **VOLUMETRIC LATTICE** dialog. *Figure 4.34* shows the application of our custom cell to a box of 10 mm by 10 mm by 10mm:

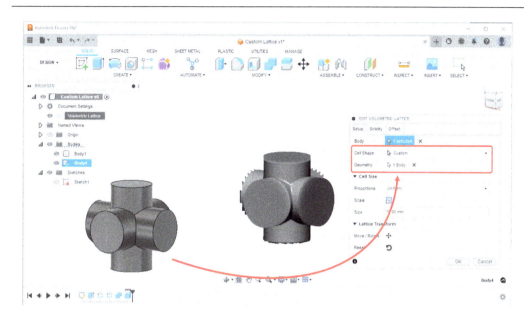

Figure 4.34 – Utilizing the solid body as a custom cell within volumetric lattice

When using a custom lattice as an input Cell Shape for a Volumetric Lattice, we have to remember that Fusion 360 is a parametric CAD tool and any changes we make to the body, which is used as input for the custom lattice, will have a downstream impact on the volumetric model, as shown in *Figure 4.35*.

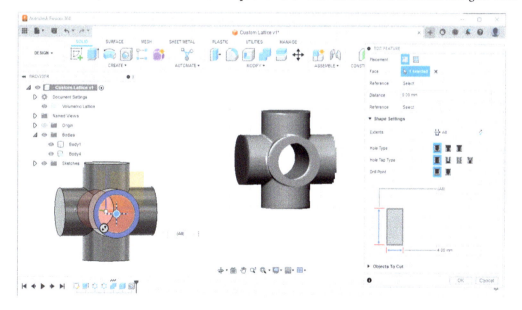

Figure 4.35 – The instant impact of modifying the custom cell solid body

Now that we've covered all the details around how to use volumetric lattice, we can apply it to the part we designed and modified in the previous subsection. You can refer to the provided file, named `Hook with drain hole and lattice`, to see the outcome of applying a gyroid lattice with a cell size of 10 millimeters and an offset apply to all 23 faces of the hook with a 5-millimeter thickness:

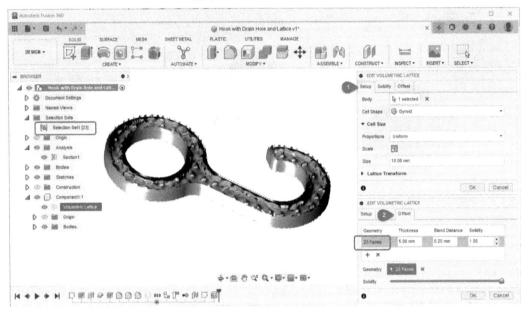

Figure 4.36 – Section view of the volumetric version of the hook model

As we previously outlined, if we wanted to export this model out of Fusion 360 to be sliced using a third-party slicer, we would have to create a mesh. After selecting the body and right-clicking, we can simply select **Create Mesh** and hit **OK** on the resulting dialog. Once we do, we will be able to visualize the meshed outcome of our model in the section view, as shown in *Figure 4.37*.

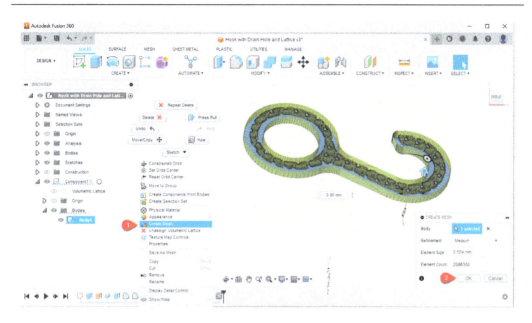

Figure 4.37 – Section view of the meshed version of the hook model

The final type of lattice we will cover in this subsection is a solid lattice. To create an example solid lattice, we will use the cross shape we have created as shown in *Figure 4.33*. Using Fusion 360's **Rectangular Pattern** tools, we can easily create a 5-by-5 matrix in the X and Y direction for this shape, as shown in *Figure 4.38*:

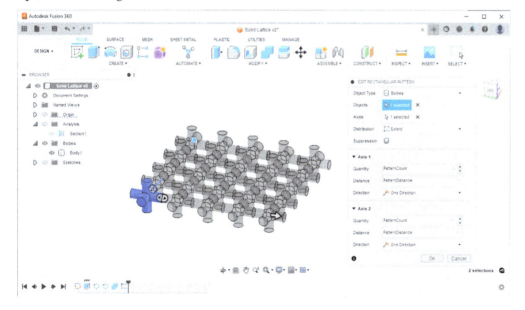

Figure 4.38 – Using a rectangular pattern on a solid body

Once we have a planar version of a solid lattice, we can simply utilize the **PATTERN ALONG A PATH** command to create multiple copies along the Z axis. Using the pattern commands and selecting a solid body as an input will result in multiple copies of that same body. In this case, as this is a 5-by-5-by-5 matrix, we will end up with 125 bodies. We can simply use the **COMBINE** command, located within the **MODIFY** panel of the **SOLID** tab, and create a single solid body to represent the lattice shape.

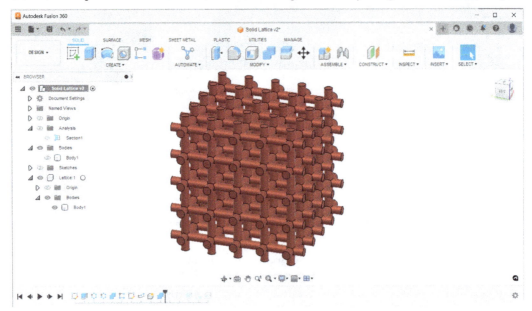

Figure 4.39 – A single solid body representing a lattice

We can then combine and intersect this lattice shape with the shelled version of the solid body we are trying to hollow out and lattice, as shown in *Figure 4.40*:

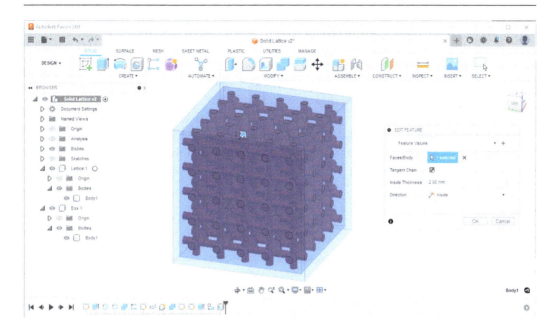

Figure 4.40 – Hollowed-out box and an internal lattice

If we apply a section analysis to the part, we can easily visualize the outcome of the hollowed-out objects and the lattice within the cavity of the part, as shown in *Figure 4.41*. Using this method, we can finalize our parts by adding the necessary drain holes with the solid modeling techniques previously discussed:

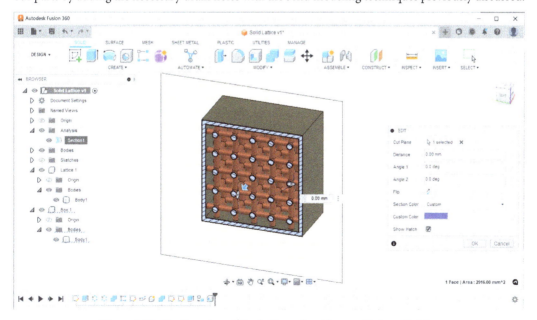

Figure 4.41 – Section view of the hollowed-out box and internal lattice

Creating lattices as solid objects is very beneficial, as designs we model as a solid geometry in parametric modeling can be easily edited later as needed. Solid lattices can also be exported as CAD files, such as STEP, SAT, and IGES. In addition, if you wanted to simulate the performance of parts with lattice structures, Fusion 360 can perform such simulations only if our models are made up of solids. Volumetric lattices cannot be simulated using Fusion 360. However, creating a solid lattice takes a lot of computing resources and can be cumbersome. Creating a lattice and shelling a given part using the **VOLUMETRIC LATTICE** dialog is more straightforward and can be used effectively to 3D print parts, if we do not need to simulate them, which is often the case for manufacturing prototype parts with resin printing.

Summary

In this chapter, we covered how to hollow solid and mesh bodies using solid modeling, mesh modeling, and volumetric modeling techniques. We demonstrated the two methods we have for hollowing a solid body. The first method was by selecting one or multiple faces and removing those faces from the solid object in order to create an outer shell of the remaining body. The second method we utilized simply carved out the inside of the solid body. We highlighted how to visualize the impact of hollowing out a solid body by changing the visualization as well as adding a section analysis.

Later in the chapter, we highlighted how to add drain holes to shelled objects to remove trapped powder or resin for 3D printing. We explained how the size and placement of the drain holes, as well as the number of drain holes needed, depend on the 3D printing technology. We demonstrated how to add drain holes for removing uncured resin from **stereolithography** (**SLA**) and **digital light processing** (**DLP**) printers. We covered the different-size drain holes required based on the viscosity of the resin and the 3D printing technology. We also highlighted where to place the drain holes so that we can effectively remove resin as well as eliminate the cupping effect, which could result in part failure during printing. We touched on how the placement of the drain holes depends on the part orientation for 3D printing.

We finished the chapter by showcasing how to add volumetric and solid lattices to our parts. One of the main reasons we want to hollow out our bodies is to lightweight them and save material and energy during 3D printing. However, not all technologies can 3D print a hollow object without needing support structures. In the final subsection of this chapter, we covered various ways we can create internal lattices, which act as support structures for printing hollowed objects with SLA and DLP printers. We talked about how hollowing out a body and adding a volumetric lattice can be accomplished with a single command within Fusion 360. We ended the chapter by talking about the differences between volumetric lattices and solid lattices and how solid lattices can be used to conduct finite element analysis simulation to see whether the 3D printed part can withstand operating conditions.

Using these techniques, we can quickly and easily cut down on the amount of material we need to 3D print parts with large cross-sections using resin or plastic powder-based printers. This takes us to the end of our discussion of design for additive manufacturing concepts.

In the next chapter, we will transition from talking about design to highlighting how to manufacture those designs by looking at how to export models created in Fusion 360 to third-party slicer software.

Part 2:
Print Preparation –
Creating an Additive Setup

In this part, we will show you how to get started with print preparation. There are two methods you can utilize during print preparation while working with models created/modified with Autodesk Fusion. First, we will highlight how to tessellate models and export them to external software, in order to slice them for 3D printing. Then, we will showcase how to prepare our models additive manufacturing using Autodesk Fusion as the slicer. We will discuss the various ways you can use the **DESIGN** and **MANUFACTURE** workspaces to aggregate your models for print preparation. Finally, we will create our first additive setup using Fusion within the **MANUFACTURE** workspace.

This part has the following chapters:

- *Chapter 5, Tessellating Models and Exporting Mesh Files to Third-Party Slicers*
- *Chapter 6, Introducing the Manufacture Workspace for Print Preparation*
- *Chapter 7, Creating Your First Additive Setup*

5
Tessellating Models and Exporting Mesh Files to Third-Party Slicers

Welcome to *Chapter 5*. In previous chapters, we talked about opening CAD files and mesh files in Fusion 360, checking for errors, and repairing them. We also covered how to edit CAD and mesh files with design for additive manufacturing principles in mind. We highlighted how to lightweight parts using automated modeling and shape optimization and simulate them to make sure that they can withstand the operating loads and boundary conditions. We also covered how to lightweight parts using hollowing and latticing methods to reduce both the amount of material and energy required. In this chapter, we will focus on taking those models out of Fusion 360 to third-party slicers.

Autodesk Fusion 360 already has a built-in slicer, and we will cover it in depth starting in *Chapter 7*. In this chapter, we will demonstrate how you can export models out of Fusion 360 into other third-party slicers for 3D printing. There are hundreds of 3D printing hardware companies on the market today. You may have heard of some of the common ones, such as UltiMaker, Prusa, Formlabs, EOS, Renishaw, Stratasys, and Nikon. Some 3D printers are open to third-party print preparation solutions and allow for software such as Fusion 360 to slice models and create toolpaths for 3D printing. Meanwhile, others require their users to utilize the proprietary slicer. Regardless of the type of hardware you use to 3D-print your models, it is a good idea to have a basic understanding of how to export models out of Fusion 360 in case you need to use them elsewhere.

Each slicer is unique and has dedicated workflows for the 3D printer it supports. Most, if not all, 3D printing slicers utilize common mesh files such as STL or OBJ file formats. More recently many free-to-use slicers such as PrusaSlicer from Prusa Research, Cura from UltiMaker, and PreForm from Formlabs have started accepting 3MF file formats.

It is less common for free-to-use 3D printing slicers to read in solid geometry from CAD software such as Fusion 360, Inventor, or SolidWorks. However, paid versions of certain slicers (e.g., Ultimaker Professional – `https://ultimaker.com/software/ultimaker-professional/` – Fusion 360, and

Autodesk Netfabb Premium –https://www.autodesk.com/products/netfabb/overview) have the capability to read in universal solid model file formats such as STEP or SAT as well as native CAD files from various CAD software.

Fusion 360 has a built-in slicer, which we will highlight in *Chapters 7* through *12*. However, Fusion 360's slicer does not support all 3D printers on the market. As we mentioned at the beginning, certain 3D printers have their proprietary slicing software. In certain organizations, software coordinators may decide which slicing software a manufacturing engineer may use, or you may simply wish to use a slicing software you are already familiar with. In such cases, after creating a model and modifying it for the 3D printing technology, which we covered how to do in *Chapters 1* through *4* of this book, you may want to export your models to your preferred slicer. In this chapter, we will go over the various methods with which you can export your Fusion 360 models to your preferred slicer.

We will start the chapter by talking about the various mesh files and how they are used for 3D printing. Next, we will focus on how to create mesh files within Fusion 360. We will end the chapter by showcasing how to export geometry as a mesh file and highlighting how you can streamline your workflow between Fusion 360 and a third-party slicer by sending your mesh files directly to your desired 3D printing utility.

In this chapter, we will cover the following main topics:

- Common mesh file formats and their differences

- Tessellation, a critical step for 3D printing with non-Fusion 360 slicers

- Exporting a mesh file or sending your models to your slicer

By the end of this chapter, you will have learned why we need to mesh certain types of models in order to 3D-print them. You will also have a better understanding of the differences between common mesh file formats and the type of information each mesh file can contain. You will be able to convert a solid or volumetric lattice body in Fusion 360 into a mesh body, and you will have a good understanding of how to export mesh bodies from Fusion 360 directly to your slicer.

Technical requirements

All the topics covered in this chapter are accessible to Startup, Educational Fusion 360 license types. If you have a commercial license of Fusion 360, in order to create a volumetric lattice and/or convert it into a mesh body, you will also need access to the *Design extension* (https://www.autodesk.com/products/fusion-360/design-extension).

Autodesk offers a 14-day trial for the Design extension. You can activate your trial from the **Extensions** dialog within Fusion 360.

Design extension access is not available for personal/hobby licenses of Fusion 360.

The lesson files for this chapter can be found here: https://github.com/PacktPublishing/3D-Printing-with-Fusion-360.

Common mesh file formats and their differences

In computer-aided design and modeling, a mesh is a collection of vertices and edges that defines the shape of an object. If we are dealing with a 2D design, we end up with a collection of lines. If we are 3D-modeling an object, we can represent a shape using polygons made up of vertices and edges. The simplest form of a polygon is a triangle. A triangle represents a face, and when we have multiple faces coming together with common edges, we can model any geometry. Whenever a model is represented using a collection of polygons, it is referred to as a mesh model. At its core, a mesh model can be represented with the coordinates of all the vertices and an identifier for each vertex. You also need a list of edges that connect two vertices. Finally, you need a list of faces that are made up of three or more edges. Because of the simplicity of the definition of a mesh file, there are tens of mesh file formats out there.

STL format

The most well-known and used mesh file format is **STL**. It is also the oldest and has the most limitations. The STL file format was created by 3D Systems to aid in its first commercial 3D printers in 1987. There are both a binary and an ASCII file format for STL. The binary version is more common as it is more compact. The ASCII version is a collection of faces and the vertices that make up those faces and can be read using a standard text editor application such as Notepad:

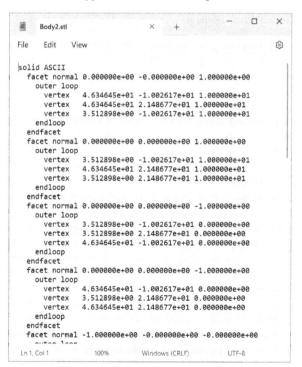

Figure 5.1 – An STL file in ASCII format as viewed in Notepad

The STL file format does not contain critical information such as units, color, or texture. Over the years, as 3D printing became more widespread and colored 3D printers became available, there emerged a growing need to have a new file format with color and texture information, the OBJ format.

OBJ format

The **OBJ** file format is the next most common file format after STL in 3D printing as it contains color and texture information. However, the OBJ file format also does not contain units even though it may include the unit information in a comment line.

3MF format

To address all the shortcomings of STL and OBJ file formats and provide a solution path for potential future improvements in the additive manufacturing industry, a consortium was created and the **3MF** file format was founded. The 3MF file format contains information about units, color, texture, material, metadata, copyright information, volumetric modeling data, beam data, slice data, as well as file security through a multitude of extensions.

There are a number of design and manufacturing software that support the 3MF file format at various levels. You can learn more about how each additive manufacturing software can either import or export 3MF files and which specific capabilities are supported in the import/export workflow in the **Compatibility** section within the **3MF Consortium** web page: https://3mf.io/compatibility-matrix/.

Fusion 360 can import 3MF files that contain unit, color, and texture data, which we covered in *Chapter 1*. Fusion 360 can export mesh data or tesselate a solid/surface model and export it as a 3MF file. We will cover the export workflow in detail in the following section.

Tessellation, a critical step for 3D printing with non-Fusion 360 slicers

As we learned in the chapter introduction, there are many slicers out there for 3D printing. Even though certain slicers can accept solid models from CAD software, more often than not, you will have to *tessellate* your models and turn them into mesh models prior to sending them to your preferred slicer. Fusion 360 can convert solid and surface models into mesh bodies using the **TESSELLATE** command located within the **MESH** tab of the **CREATE** panel.

Tessellating a solid body

The **TESSELLATE** command has multiple options to control what the mesh outcome will look like. After selecting a body or a component to tessellate, you can use the **Refinement Settings** dropdown and change the quality of the tessellated mesh body. The default refinement option is **medium**, but I would recommend using **high** in order to get a more refined mesh outcome, especially around a curved surface. The refinement option you choose controls the input values for **Surface Deviation** and **Normal Deviation** based on the selected geometry. In addition, you can reduce the input values for **Maximum Edge Length**, which would create smaller lines, and **Aspect Ratio**, which would create a mesh body with more uniform-looking triangles. The **Preview** checkbox within the tessellate dialog gives you a visual representation of the meshed outcome, as well as a numerical count for the number of triangles the mesh body will have.

Figure 5.2 shows a side-by-side visualization of a solid body and a preview of what a mesh version of that same body would look like:

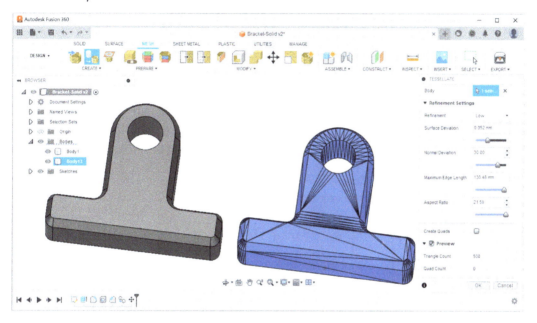

Figure 5.2 – Tessellating a solid body to convert it into a mesh body

Once a solid body has been tessellated, it will be consumed in the new mesh body, which will also be marked with a new icon within the browser. When tessellating solid bodies, every face of the solid body will be a unique face group within the mesh body. If you want to visualize the face groups of a mesh body, you can toggle the **Display Mesh Face Groups** option within the **INSPECT** panel, as shown in *Figure 5.3*.

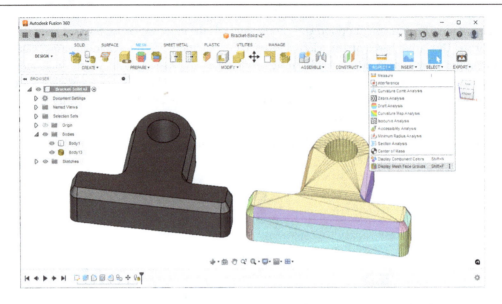

Figure 5.3 – Mesh face groups can be visualized after tessellating a solid body

Another example of why we need to tessellate a body prior to exporting to a third-party slicer can be demonstrated for bodies with a volumetric lattice definition, which we'll see in the next section.

Tessellating a body with a volumetric lattice

As you may recall, we covered volumetric lattices in *Chapter 4*. *Figure 5.4* shows the bracket we have worked on in this chapter, with a gyroid cell shape lattice and a skin around most of the exterior faces.

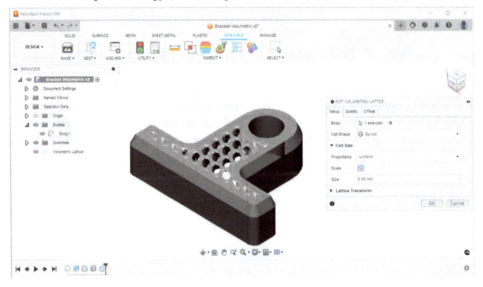

Figure 5.4 – Bracket with a volumetric lattice

When attempting to export models created in Fusion 360 with a volumetric lattice using traditional export methods, such as the **3D PRINT** command located within the **UTILITIES** tab of the **MAKE** panel, we can see that Fusion 360 cannot generate a mesh outcome as a part of the export workflow. As shown in *Figure 5.5*, a mesh preview is not available, and the number of triangles is 0 when the input body contains a volumetric lattice.

Figure 5.5 – Attempting to export a body with a volumetric lattice

In such cases, we need to convert our volumetric model into a tessellated version prior to export. Fusion 360 has a dedicated workflow for converting bodies with volumetric lattice data into meshes. We cannot use the **TESSELLATE** command located within the **MESH** tab of the **CREATE** panel.

In order to tessellate a body with volumetric lattice data, right-click on the desired body in the **BROWSER** window and select **Create Mesh** from the pop-up menu, as shown in *Figure 5.6*. Then, we can edit the **Refinement** option, and upon selecting **OK**, Fusion 360 will create a new mesh body and hide the original solid body with the volumetric lattice.

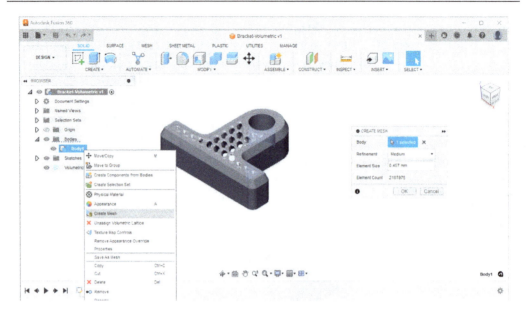

Figure 5.6 - Selecting a volumetric lattice body and creating a mesh body

Once the mesh body is created, we can simply select it from the **BROWSER** window, right-click and select **PROPERTIES** from the pop-out menu, and observe information such as **Facet Count**, **Vertex Count**, and **Center of Mass**. We can also get information about the quality of the mesh, such as whether it encloses volume and whether all the triangles are oriented properly, as shown in *Figure 5.7*.

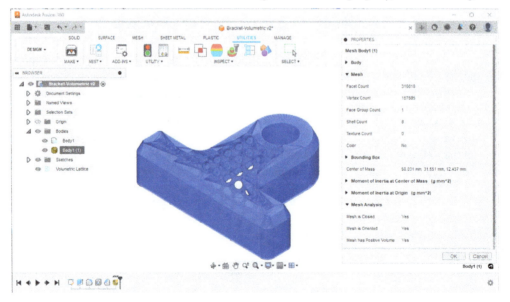

Figure 5.7 – Properties of a mesh body created from a volumetric lattice body

In this section, we covered how to tessellate solid bodies as well as bodies with volumetric lattice data. We touched on the need to tesselate bodies before exporting them to third-party slicers, as models with volumetric lattice data cannot be exported to third-party slicers directly as 3MF files. In the next section, we will highlight how to send solid and mesh data to third-party slicers.

Exporting mesh files or sending models to your slicer

Transferring a model from Fusion 360 to a third-party slicer from the design workspace has a very similar workflow to tessellating models. Instead of using the **MESH** tab to tessellate the model, we can simply use the **3D PRINT** command located within the **UTILITY** tab of the **MAKE** panel. The **3D PRINT** dialog allows us to choose a single object to be exported out of Fusion 360. That selection could be a solid body, a surface body or a mesh body, or a component containing one or more of those bodies.

Exporting models directly to third-party slicers

As shown in *Figure 5.8*, by selecting a body and choosing the relevant refinement options, we can preview the mesh outcome within the **3D PRINT** dialog. Just like in the **TESSELLATE** command, we can edit **Refinement Options** to get the desired quality out of our export process.

The **3D PRINT** dialog has several unique options, which we will cover next. The first one is the **Format** dropdown. Using the **Format** menu, we can choose to export our models as STL, OBJ, or 3MF files. If we choose an STL or OBJ file, we also have the ability to scale our model during the export process.

Models created with Fusion 360 always include units. If we have a model with a certain edge with a dimension of 1,000 millimeters, upon exporting it in an STL or OBJ file format with the unit type of millimeters, we will end up with a mesh file that will represent that same edge as 1,000, as STL and OBJ files do not have a unit system. If we change the unit type to centimeters within the **3D PRINT** dialog during the export process, the resulting file will list the same edge as 100. This means Fusion 360 will scale the model by a factor of 0.1 when we choose centimeters as the unit type.

> **Note**
> The following scale factors apply when exporting models as STL or OBJ files using the **3D PRINT** dialog: millimeter = 1, centimeter = 0.1, meter = 0.001, inch = 0.03937, and foot = 0.003281.

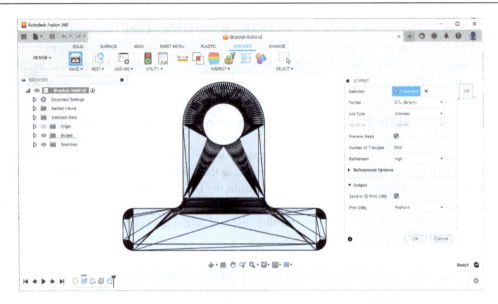

Figure 5.8 – The 3D PRINT dialog to export a model as an STL file

The **3D PRINT** dialog can also automatically launch a print utility and open the newly generated STL, OBJ, or 3MF file within that software. *Figure 5.8* shows the example of an STL file being created and sent to the PreForm software from Formlabs. After selecting **OK** in the **3D PRINT** dialog, PreForm will launch, and the STL file will be loaded into the last used printer setup as a new job, as shown in *Figure 5.9.*

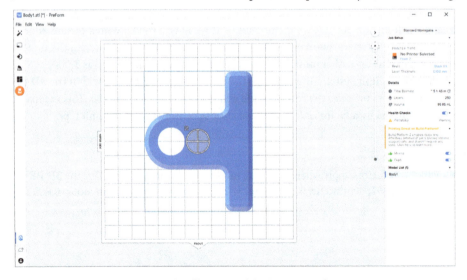

Figure 5.9 – STL file in Formlabs PreForm software

When using the **3D PRINT** dialog, we are not limited to selecting single bodies. We can also choose an entire assembly as an input, and export using the STL, OBJ, or 3MF format, as shown in *Figure 5.10.*

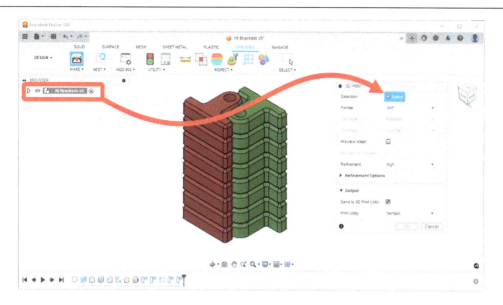

Figure 5.10 – Parts arranged in 3D as an assembly being exported to Autodesk Netfabb

In *Figure 5.10*, we have chosen to transfer all 16 components to a 3D print utility named Autodesk Netfabb using 3MF as the file format. As shown in *Figure 5.11*, once Autodesk Netfabb launches and all parts are loaded into its browser, we can see that the names of all 16 parts have come over to Netfabb. The part arrangement in Fusion 360 is also transferred to Netfabb with the same command. This is because we have chosen 3MF as the export file format, which contains all this information and more.

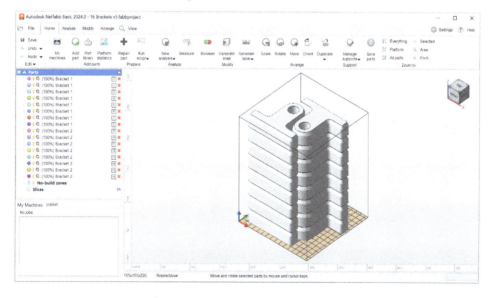

Figure 5.11 – Component names and part arrangement in Netfabb

In the *Tessellating a body with a volumetric lattice* subsection, we demonstrated how to convert a body with volumetric lattice data into a tessellated mesh object. *Figure 5.12* shows that same mesh body being transferred as a 3MF file into a third-party 3D print utility called PreForm. Whenever we are exporting a mesh body or transferring it to a 3D print utility, the **3D PRINT** dialog will automatically hide the irrelevant options, such as **Preview Mesh**, **Number of Triangles**, and **Refinement**, as those settings are only applicable when exporting a solid body as a mesh object.

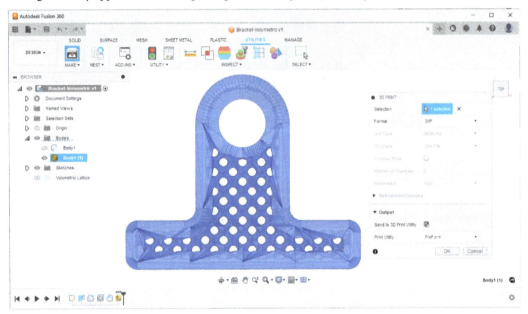

Figure 5.12 – A tessellated mesh body being used as an input to export to PreForm

Regardless of whether we are sending a solid body or a mesh body to a 3D print utility, the outcome will be very similar. In the example in *Figure 5.12*, where we sent a mesh body as a 3MF file to PreForm, the Formlabs PreForm software will start and the 3MF file will open within the application, as shown in *Figure 5.13*:

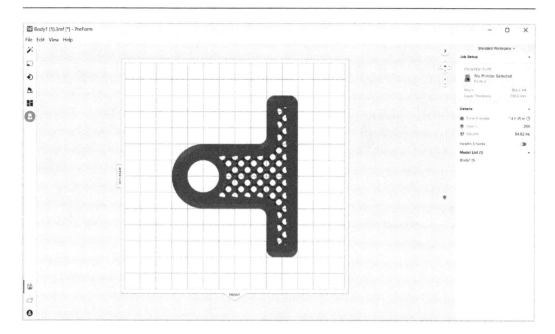

Figure 5.13 – A 3MF file sent from Fusion 360 to PreForm

The **3D PRINT** dialog output options for the print utility will vary based on the software you have installed on your computer.

Exporting models directly to non-Fusion 360 slicers

So far, we have demonstrated how models can be sent to PreForm and Netfabb, which are both explicitly available options, assuming that you have those software options installed, for the **Print Utility** drop-down menu. If you are using a slicer that is not available within the **Print Utility** dropdown, you can choose the **Custom** option and point to the executable for the slicer of your choice within the **Open** dialog accessible after selecting the folder icon next to the **Application** option. *Figure 5.14* shows how to select a custom print utility and assign `prusa-slicer.exe` as the application.

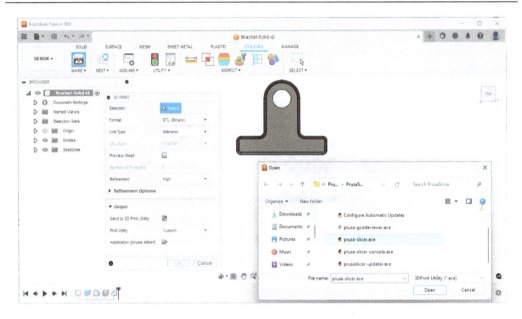

Figure 5.14 – Selecting a custom print utility executable as a part of the 3D PRINT dialog

After designating the application as PrusaSlicer and selecting **OK** within the **3D PRINT** dialog, we can transfer our solid or mesh bodies to this slicer as STL, OBJ, or 3MF files. *Figure 5.15* shows the outcome of transferring an STL file from Fusion 360 to PrusaSlicer using the **3D PRINT** dialog.

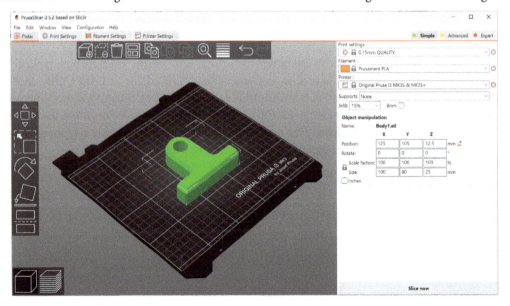

Figure 5.15 – An STL file sent from Fusion 360 to PrusaSlicer

In certain cases, we may not want to send our models to a **3D print utility** but rather export them as a mesh file. *Figure 5.16* shows how we can utilize the **3D PRINT** dialog to export our models out of Fusion 360 as STL files. To achieve this, we simply need to uncheck the **Send to 3D Print Utility** checkbox and select **OK**.

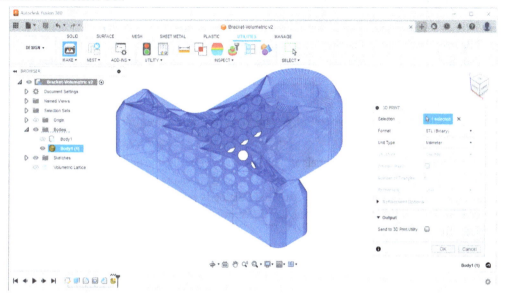

Figure 5.16 – Exporting an STL file out of Fusion 360

After pressing **OK** within the **3D PRINT** dialog, we will see the **Save As** dialog, as shown in *Figure 5.17*.

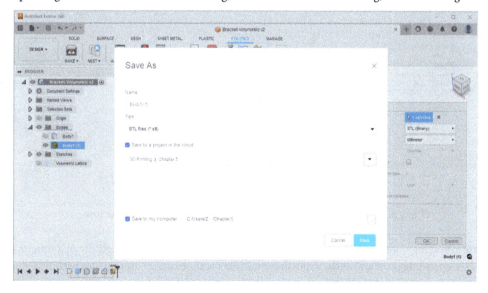

Figure 5.17 – Saving the STL file to your local hard disk and your project

Using this standard dialog, we can export all our files as STL, OBJ, or 3MF files to our computer and/ or to a **Project** within our **Fusion Team**.

Exporting models out of Fusion 360

Fusion 360 has several other options for exporting models as mesh bodies. One such option is to use the **File** dropdown and utilize the **Export** command. Once the **Export** dialog is visible, you can change your file extension to your desired mesh type, and Fusion 360 will export your model as such.

> **Note**
> If you are using the **File | Export** option, you do not have control over the mesh quality.

My preferred method for exporting models as mesh files is by selecting the body or component within the browser or canvas, right-clicking on the selection, and choosing the **Save As Mesh** option, as shown in *Figure 5.18*. The resulting **SAVE AS MESH** dialog is very similar to the **3D PRINT** dialog we covered throughout this section with one key difference. The **SAVE AS MESH** dialog also allows you to select the export **Structure** setting.

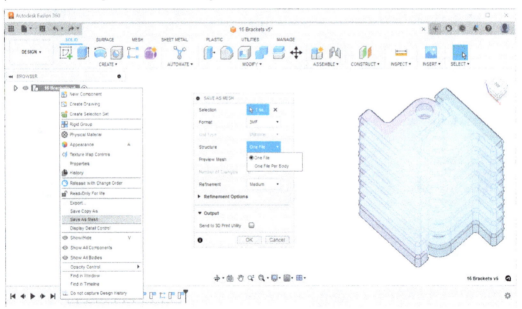

Figure 5.18 – The SAVE AS MESH command can control the file structure

If you are exporting multiple components, you can choose to save them as a single file or a unique file per body. Depending on your needs, exporting multiple components as unique files with this single dialog can save you a lot of time and automate your file export workflow.

Summary

In this chapter, we talked about how a body can be represented as a mesh object, and the generic file specifications for a mesh file format. We introduced the STL file format and talked about its limitations with respect to units, colors, and textures. We briefly touched on how the OBJ file format includes color and texture information but lacks information about units. We introduced the 3MF file format and referenced the various extensions the 3MF file format has for defining not only geometry and units but also color, texture, beam definition, and volumetric data. We also talked about how to tessellate both solid bodies and bodies with volumetric lattice data within Fusion 360. We ended the chapter by highlighting various ways we can export geometry out of Fusion 360 as mesh data.

We demonstrated how to export models directly to third-party slicers, such as Formlabs PreForm and Autodesk Netfabb, as well as other slicers, such as PrusaSlicer, which are not necessarily in the native export option list within Fusion 360's **3D PRINT** dialog. Finally, we demonstrated how to export models out of Fusion 360 as a single file or one file per body to automate our export process.

Using these techniques, we can easily export our models as mesh bodies to our preferred slicers so that we can 3D print our parts quickly.

In the next chapter, we will transition from talking about third-party slicers and start introducing the MANUFACTURE workspace of Fusion 360 for print preparation in order to use this workspace as a native slicer.

6

Introducing the Manufacture Workspace for Print Preparation

Welcome to *Chapter 6*. In previous chapters, we talked about working with various file types using Fusion 360. We showed you how to inspect and repair problematic mesh models. We also touched on how to design parts with additive manufacturing principles in mind. We covered how to use lightweight parts using various techniques such as latticing, and we simulated them to make sure they would perform under operating conditions. We also covered how to tessellate our models so that we can export them to other slicers. In this chapter, we will focus on how to use Fusion 360 more effectively to manipulate models for print preparation.

Until now, most of the models we've worked with were single parts. There are many reasons why we may want to make modifications to a part before slicing it for 3D printing. We may want to scale up or scale down our part to print a different size. We may want to orient and arrange our parts on the build platform and print multiple copies of the same part in one print. As we covered in the previous chapters, we may want to hollow and lattice our parts based on their orientation and add drain holes. We may want to add or modify certain features so that our parts can be printed without support structures or adhere to the build plate more effectively. Similarly, if we are working with an assembly, we may want to only print a select few parts within that assembly. Naturally, the positioning and orientation of those parts will differ on the 3D printer build plate from how they were originally designed and oriented. For these reasons and more, we need a new version of our design for 3D printing. In *Chapter 5*, after exporting our model to a third-party slicer, the file we created and saved within our slicer was that new version. In this chapter, we will explore the various ways Fusion 360 enables us to create a new version of our design specifically for 3D printing.

The main benefit of staying within Fusion 360 to create a new version of our design for manufacturing is that Fusion 360 keeps the associativity between the as-designed version of a given model and the to-be-manufactured version of the same part. This means that if we decide to make a change to the original design upstream, our downstream manufacturing model will automatically be updated accordingly.

There are two ways Fusion 360 allows you to create a manufacturing model while keeping associativity. The first method is to create an entirely new Fusion 360 design and utilize the insert functionality to import components from other designs. The second method is to utilize the Manufacturing Model functionality in the **MANUFACTURE** workspace. In this chapter, we will go over both workflows in detail and highlight the benefits and drawbacks of each method while showcasing common examples of design changes we may want to make before 3D printing.

In this chapter, we will cover the following topics:

- Using the derive workflow to manage model changes for print preparation
- Creating manufacturing models for 3D printing
- Common part modifications for 3D printing

By the end of the first section, you will have learned how to utilize Fusion 360 to create new models and gather components from other designs to prepare them for 3D printing. In the next section, you will learn about how to access the **MANUFACTURE** workspace and create manufacturing models. By the end of this chapter, you will have mastered how to make design changes upstream in your original design documents and how they propagate down to your manufacturing models. You will also understand how the changes you make downstream to the manufacturing model do not flow upstream to the original design.

Technical requirements

Most of the topics covered in this chapter are accessible to Personal, Trial, Commercial, Startup, and Educational Fusion 360 license types. There are certain Fusion 360 commands, such as **ARRANGE**, that are not accessible with the Personal license type. In examples where a certain function is not accessible with a Personal license type, an alternative feature will be presented so that you can continue with the exercise.

If you have a commercial license of Fusion 360, to create a volumetric lattice, you will also need access to the Design extension (`https://www.autodesk.com/products/Fusion-360/design-extension`).

Autodesk offers a 14-day trial for the Design extension. You can activate your trial within the **EXTENSIONS** dialogue of Fusion 360.

Design extension access is not available for personal/hobby licenses of Fusion 360.

The lesson files for this chapter can be found here: `https://github.com/PacktPublishing/3D-Printing-with-Fusion-360`.

Using the derive workflow to manage model changes for print preparation

If you have used a third-party slicer (for example, Cura, PrusaSlicer, or PreForm), you are probably familiar with how, upon launching such software, they often give you an empty workspace where you can insert your parts to be 3D-printed. Whenever we want to use Fusion 360 as a slicer, we can follow a similar workflow by creating a new design and saving it with a new name. Once we have saved our design, we can use it as an empty workspace to aggregate bodies and components from other projects or folders within our Fusion Team to prepare them for 3D printing. *Figure 6.1* shows how to utilize the **Insert Derive** functionality located in the **SOLID** tab's **INSERT** panel after saving our Fusion 360 design with the name `Example A`:

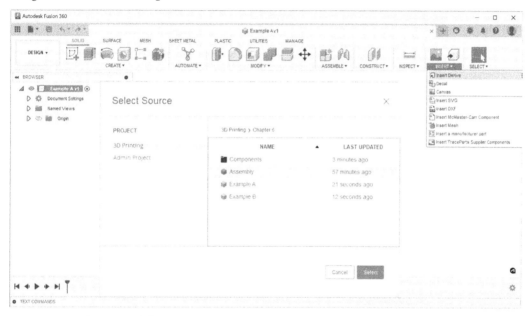

Figure 6.1 – Selecting a source model to insert into a saved Fusion 360 design

To demonstrate this functionality, we will be utilizing a Fusion 360 design named `Assembly`. After selecting the source model, Fusion 360 will open the model in a new tab and display the **DERIVE** dialogue, as shown in *Figure 6.2*. At this point, we can select which objects we would like to derive from the source file to our destination file. In this case, we will select multiple components using the browser, as shown in the same figure. Alternatively, we can also choose to use the canvas. Keep in mind that the default selection method Fusion 360 utilizes will select bodies instead of components within the **DERIVE** dialogue. If you want to select a component, you can do so by holding and clicking the left mouse, which will bring up a secondary dialogue with two tabs called **Depth** and **Parents**, as shown in *Figure 6.2*. If you go to the **Parents** tab, you will be able to select the component that the body you are selecting belongs to and add it to the **DERIVE** dialogue. The **DERIVE** dialogue has a couple of additional options we should highlight. The first one is the **Place Objects At Origin** checkbox. After selecting this option and completing the derive function, Fusion 360 will move all selected components to the global origin of the new Fusion 360 design document. This is very useful since a given part in an assembly could be far away from the origin, and upon inserting it into a new document, you may not be able to visualize it easily if it is hundreds or thousands of millimeters away from other components. The next useful option within the **DERIVE** dialogue is the ability to bring in the parameters that make up the derived components. *Figure 6.2* shows that we can choose either **Favorites** or all the parameters that make up the component upon deriving it within the new Fusion 360 design:

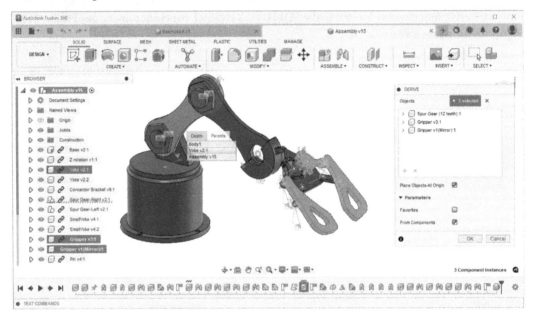

Figure 6.2 – Selecting multiple components from the source assembly

After completing all our selections and pressing the **OK** button within the **DERIVE** dialogue, Fusion 360 will close the source model and display all the inserted components in our destination design, named `Example A`, as shown in *Figure 6.3*. Notice how all the components have been inserted using a single Derive function in the timeline. You will also notice that all four components are on top of each other as we have used the **Place Objects At Origin** option within the **DERIVE** command. This moved all the components near the center of the new Fusion 360 design. Using **Display Component Colors** within the **SOLID** tab, which can be found in the **INSPECT** panel, we can give each component a unique color:

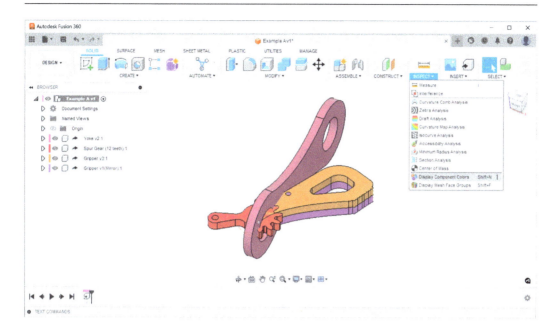

Figure 6.3 – Inserting multiple components from other designs into a new design

If needed, we could repeat the **Insert Derive** command and add components from other Fusion 360 design documents to this design. Once we are done with the insert workflow, all the components will likely intersect each other if we use the **Place Object At Origin** option within the **DERIVE** dialogue. To organize all inserted components, we could utilize the **ARRANGE**, **ALIGN**, or **MOVE** commands located in the **SOLID** tab's **MODIFY** panel.

> **Important note**
>
> Please note that the **ARRANGE** feature is not available for Personal/Free licenses of Fusion 360. If you have a Personal/Free license, you can orient and place components using the **ALIGN** or **MOVE** commands instead of the **ARRANGE** command.

Figure 6.4 shows the outcome of using the **ARRANGE** command, and nesting all four components on the *XY* plane, for a virtual 3D printer platform that is 210 mm long and 210 mm wide, while keeping the components 10 mm apart from each other and 10 mm away from the edges of the build plate. The **ARRANGE** command also automatically orients the selected components so that the largest face of a given component is facing the build plate. In this instance, we have chosen the components using the browser, so we can have Fusion 360 auto-orient the components. Alternatively, we can choose a component using the canvas by selecting a specific planar face within the **ARRANGE** command. When using this method, the **ARRANGE** command will reorient the part if needed and have the selected surface face up or face down, depending on our choice for the **Object Selected by Face** option:

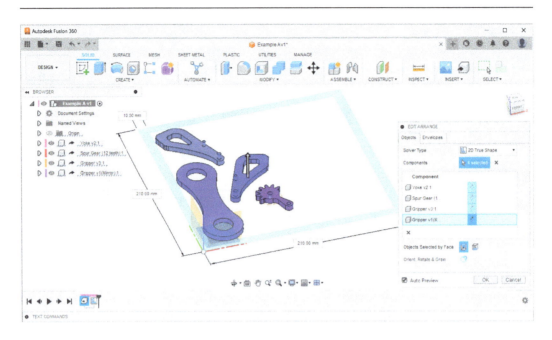

Figure 6.4 – Arranging components on the XY plane

Now that we have our four components arranged on the *XY* plane with the orientation we desire, we can talk a little bit about how to modify these components, as well as add new components to our build. To make the visualization a little bit easier, we can introduce a new sketch into the *XY* plane and build a rectangle that represents our build plate. In this example, since we are organizing our components for a printer with a build plate 210 millimeters wide and 210 millimeters long, we can utilize the sketch tools in Fusion 360 to make such a rectangle and have it be visible to help us when we are adding new parts or modifying existing ones. One common part modification for 3D printing is to utilize the **SCALE** command. The **SCALE** command is located within the **SOLID** tab's **MODIFY** panel and allows us to select bodies or components and uniformly or non-uniformly scale them up or down. *Figure 6.5* shows a preview of three parts that have been scaled down to 75% of their original size. As we don't need to scale these parts, we will click **Cancel** within the **SCALE** dialog and proceed to the next step:

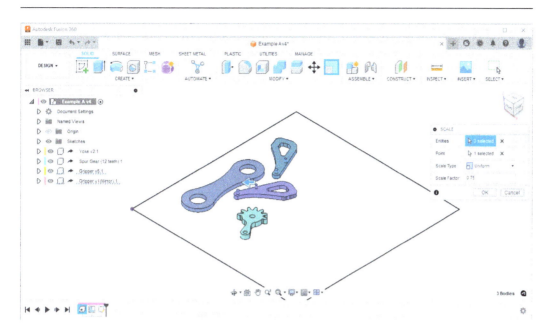

Figure 6.5 – Scaling components within the 3D printer build plate context

Another common modification before 3D printing is hollowing and latticing. *Figure 6.6* shows how we can simply select the **SHELL** command located within the **SOLID** tab's **MODIFY** panel to select two faces of the component named `Yoke` to create a shelled-out version:

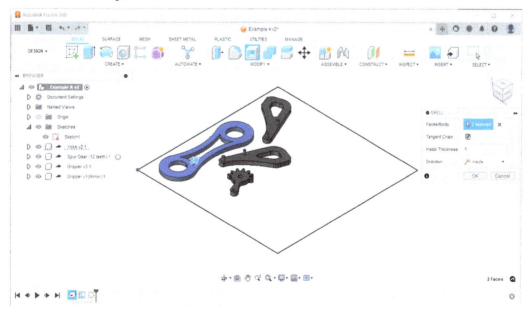

Figure 6.6 – Modifying components within the 3D printer build plate context

Another example where we can benefit from the design workspace of Fusion 360 for a setup such as this one is to select a given component and create duplicates so that we can print multiples of the same component. *Figure 6.7* shows how we can select the part named Spur Gear and create a 3x3 rectangular pattern along the *X* and *Y* axes for a total of nine components and suppress the one which would interfere with another part, to create eight total Spur Gear components:

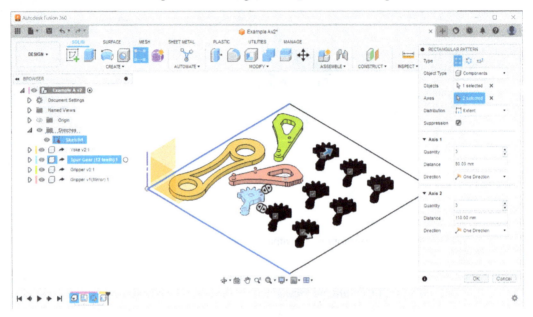

Figure 6.7 – Creating a rectangular pattern within the 3D printer build plate context

The main benefit of using a new Fusion 360 design document to make changes such as orienting parts, scaling them, or hollowing them out is the fact that such changes only impact the document that we are working on and not the source component. One way to better understand this point is to look at the document named Assembly. As we can see in *Figure 6.8*, our source Fusion 360 design document has not changed, even though we utilized the **SHELL** command on the Yoke component and created seven copies of the Spur Gear component within our Example A manufacturing document:

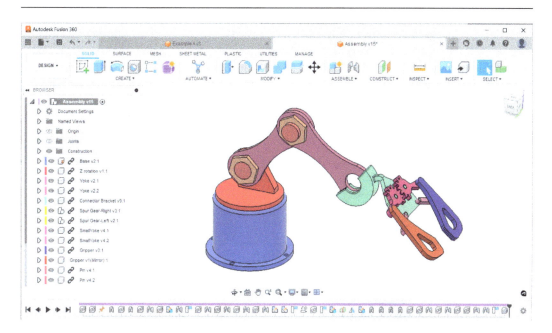

Figure 6.8 – The original assembly is unchanged

If anything changes in the design criteria, a designer can still change the original design. They can do this within the Fusion 360 Design document named Assembly or within the source component itself. *Figure 6.9* shows how you can select several features from the timeline and suppress them to eliminate a hole within the Gripper source component:

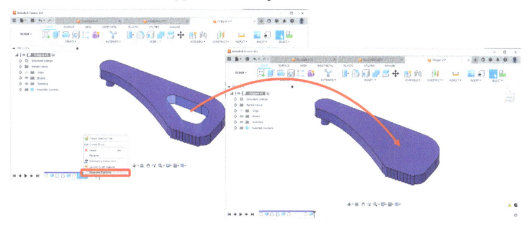

Figure 6.9 – Suppressing features to modify the source component

After saving this change to the source component, if we switch tabs and look at the `Assembly` Fusion design, we will see a warning message that will inform us that one of the source components that makes up this assembly has been updated. Upon pressing the **Get latest** command using the browser, our assembly model will update, as shown in *Figure 6.10*, and the latest version of the `Gripper` component will be visible without holes:

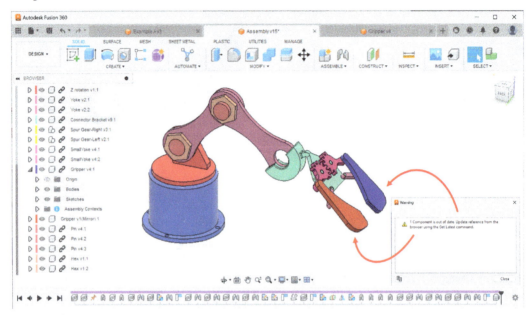

Figure 6.10 – A warning message indicating an update to a source component

After saving the latest version of the source component in the assembly document, if we switch our focus to the `Example A` Fusion design, we will see five warning signs, as shown in *Figure 6.11*. These warning signs inform us that one or more components in this Fusion 360 design are out of date and that we should get the latest versions of them:

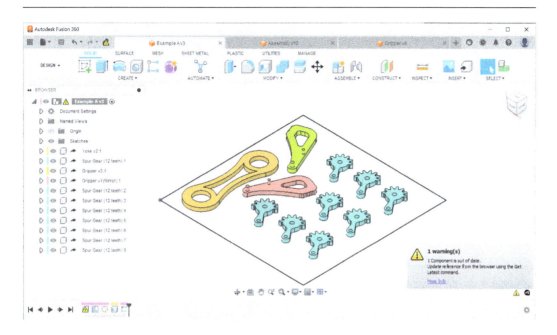

Figure 6.11 – Warning signs indicate that a component is out of date

After we update the reference, by double-clicking the warning icon in the browser, all the warning signs will disappear, and both instances of the Gripper component will be updated so that they are solid objects with no holes, as shown in *Figure 6.12*:

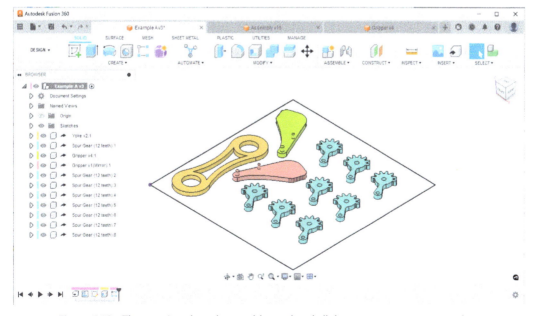

Figure 6.12 – The warnings have been addressed and all the components are up to date

In this section, we talked about how we can use an empty Fusion 360 design document as a manufacturing model, where we can insert components from other Fusion 360 designs and reorient and arrange them using the functionality available within the **MODIFY** panel of Fusion 360. We also demonstrated how changes we make to the manufacturing model do not impact the source components. However, any changes we make to a source component will flow downstream and have an impact on our manufacturing model. In the next section, we will switch from the design workspace to the manufacturing workspace of Fusion 360 and create new components and edit existing ones in that workspace.

Creating Manufacturing Models for 3D Printing

In the previous section, we talked about how to utilize the **DERIVE** functionality to aggregate components from different sources into a new Fusion 360 design document and manage those components for 3D printing. In this section, we will touch on how to prepare models for 3D printing within the same Fusion 360 design document.

In *Chapter 3*, we demonstrated how to change from the **DESIGN** workspace to the **SIMULATION** workspace using the workspace switcher menu. In this chapter, we will switch back and forth between the **DESIGN** and **MANUFACTURE** workspaces. To show you how to create a new manufacturing model within the same Fusion 360 design document, we will be using a new design document named Example B. As shown in *Figure 6.13*, we can access the workspace switcher by clicking on the **DESIGN** drop-down menu in the top-left corner of Fusion 360. In this section, we will be utilizing the **MANUFACTURE** workspace:

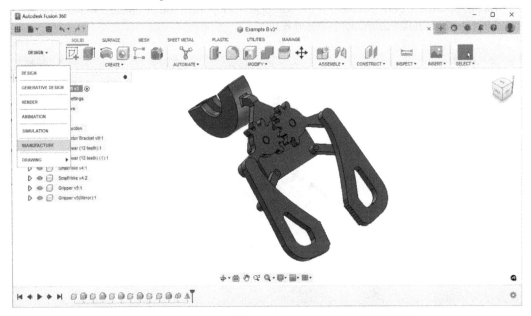

Figure 6.13 – Switching from the DESIGN workspace to the MANUFACTURE workspace

Once we are in the **MANUFACTURE** workspace, we will see a new set of ribbon tabs for different manufacturing setups, such as **MILLING**, **TURNING**, **ADDITIVE**, **INSPECTION**, and **FABRICATION**. In this chapter, we will be focusing on the **ADDITIVE** tab. After selecting the **ADDITIVE** tab, we will also see multiple new panels in the ribbon. Changing the workspace from **DESIGN** to **MANUFACTURE** automatically creates a new version of the Fusion 360 design, similar to how the **Insert Derive** functionality creates a new version within a different design document. This new driven copy can be found in the browser under the **MODELS** tab. Any changes we make to a model within the **MANUFACTURE** workspace, such as moving its position within the context of the build volume of our 3D printer, will not affect the model within the **DESIGN** workspace. However, we are limited in what changes we can make to the driven model. That is why we may want to create a new manufacturing model that will allow us to edit the geometry similar to how we can modify it within the **DESIGN** workspace.

One key difference between using the manufacturing model in the **MANUFACTURE** workspace versus a manufacturing model we can create ourselves using the **Insert Derive** functionality we covered in the previous section is that the **Insert Derive** functionality allows us to bring the CAD parameters into our new design. The manufacturing model we create in the **MANUFACTURE** workspace does not carry along the model parameters. Therefore, we will not be able to edit our model using the parameters with which it was built. When we are editing our manufacturing model in the **MANUFACTURE** workspace, we will have to create new sketches and features that are independent of the features and parameters that were used to create the original design.

Within the **SETUP** panel of the **ADDITIVE** tab, we can create a new manufacturing model, as shown in *Figure 6.14*. This action creates a new manufacturing model named `Manufacturing Model 1` and displays it under the manufacturing models list within the browser:

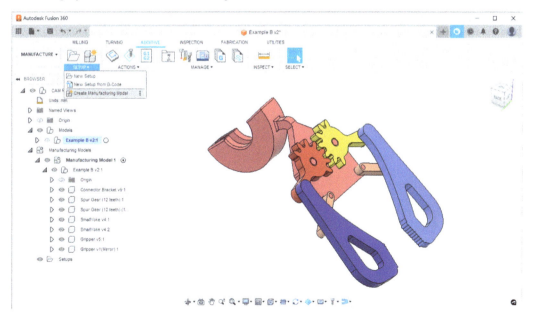

Figure 6.14 – Creating a new manufacturing model within the MANUFACTURE workspace

In the next chapter, we'll create an additive setup so that we can arrange and orient our parts within the context of our 3D printer. In this chapter, we will focus on editing the manufacturing model to make slight modifications to it. To modify components within our design, we can simply right-click on the manufacturing model and select **Edit Manufacturing Model**, as shown in *Figure 6.15*.

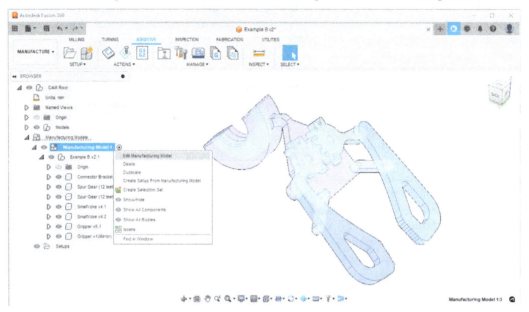

Figure 6.15 – Editing the manufacturing model

When we are editing a manufacturing model, we will still be in the **MANUFACTURE** workspace of Fusion 360. Yet Fusion 360's ribbon will update, and we will see tabs we are used to seeing in the **DESIGN** workspace of Fusion 360, such as **SOLID**, **SURFACE**, **MESH**, and **PLASTIC**. The browser will also be customized for the items related to the manufacturing model and will not display any additive or subtractive setup listed under the Setups folder. As shown in *Figure 6.16*, we can create sketches and profiles to add or remove material from our original design:

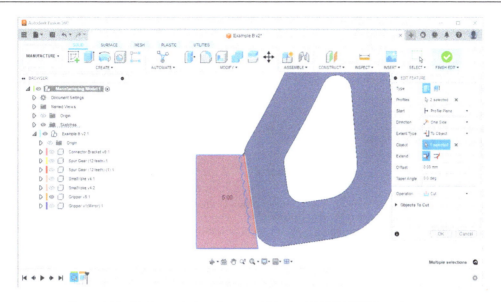

Figure 6.16 – Making a design change within the MANUFACTURE workspace

In this example, we have chosen to remove the semicircular pattern from the `Gripper v5` component by sketching a profile and using the **Extrude: Cut** operation on the source body. As you can see in *Figure 6.14*, even though the `Gripper v5` and `Gripper v1 (Mirror)` components were created in the **DESIGN** workspace using the **MIRROR** feature, updating one of them does not automatically update the other as the mirrored component is not a second instance of the original component; rather, it is an independent component:

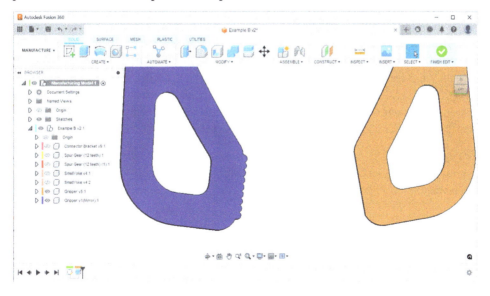

Figure 6.17 – Impact of the design change on mirrored components

If we switch back to the **DESIGN** workspace, as shown in *Figure 6.18*, we will not see any change to the source components. Changes we make to the manufacturing model within the **MANUFACTURE** workspace only impact that specific manufacturing model and do not affect the original design or any other subsequent manufacturing models:

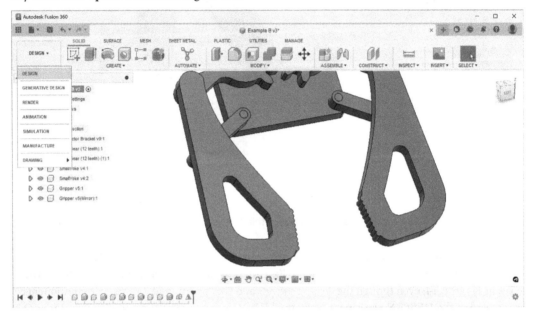

Figure 6.18 – Impact of the edited manufacturing model on the design

We can use this to our benefit and create multiple manufacturing models and have different versions of the same model saved within a single Fusion 360 design document. A common reason to use this workflow is to create a unique version of a component for each manufacturing method you may need to utilize in a hybrid manufacturing scenario. In certain cases, we may want to 3D-print our model with a certain set of features, inspect the outcome, and use subtractive operations such as surface finishing or drilling to finish our manufacturing process to achieve our desired outcome. In such cases, we may create **Manufacturing Model 1** for our additive manufacturing setup, **Manufacturing Model 2** for our subtractive manufacturing setup, and **Manufacturing Model 3** for our inspection setup.

Another key difference between using the **DERIVE** workflow versus creating a manufacturing model within the **MANUFACTURE** workspace and editing models within that workspace is around data management:

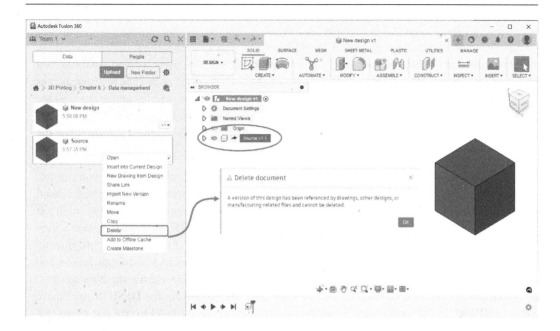

Figure 6.19 – Deleting a document that's used in other designs

If we have a model in the **DESIGN** workspace, after creating one or multiple manufacturing models and editing them, we can delete the original Fusion 360 design document, and all related manufacturing data will be deleted.

If we utilize the **DERIVE** functionality and insert components from other Fusion 360 designs into a new Fusion 360 design document, we cannot delete the source document. To delete the source document, we have a handful of options. First, we can delete the new design document and then delete the source document. Alternatively, within the new design, we can break the link between the source document and the new design, and then delete the source document, assuming that the new design document does not have previously saved versions containing a reference to the source document. Our last option is to move the source document from its current Fusion 360 project to a new project where we collect and manage all items we wish to delete in the future. Once we are ready to delete everything within that project, we can archive it using Fusion's **TEAM** functionality and then delete it. If we do not utilize one of these three options, we will be blocked while attempting to delete the source Fusion 360 design document and see a warning message, informing us about the fact that the source document is referenced in other Fusion 360 designs, as shown in *Figure 6.19*.

Common part modifications for 3D Printing

There are many reasons why you may want to modify a model for 3D printing. If you have a part that is too large for your 3D printer's build volume, you may need to cut it into multiple smaller pieces so that you can manufacture it. You may also want to add pins and holes to each piece so that the cutout parts can easily align and fit together after printing. You may want to change the dimensions of certain features, such as holes or pockets, to add small tolerances so that the surfaces where these features will mate with other objects have the desired clearance once the parts are manufactured. You may need to remove small features your printer may not be able to print or include features such as fillets to smooth out sharp edges. You may rearrange and orient your parts to minimize the need for support structures during 3D printing. Once your parts are in their desired orientation, you may want to hollow them out for lightweighting and include internal lattices to add stiffness and eliminate the need to create support structures inside the hollowed-out body.

In this section, we will highlight a few of these common part modification methods you can use in Fusion 360. As we covered in the previous sections, you can make your manufacturing-related part modification in one of two ways: you can create a new Fusion design and insert the model(s) you wish to modify, or you can switch over to the **MANUFACTURE** workspace, create a new manufacturing model, and edit it. In this section, we will demonstrate all the modifications directly within the **MANUFACTURE** workspace so that we can keep all the changes in a single Fusion 360 document. To demonstrate a few of the common part modifications, we will use the `Example C` part, which can be seen in *Figure 6.20*. As we can see, after switching from the **DESIGN** workspace to the **MANUFACTURE** workspace, we will create a manufacturing model to edit our geometry. In this example, we will create four manufacturing models using the **Create Manufacturing Model** command, which is located within the **ADDITIVE** tab of the **SETUP** panel:

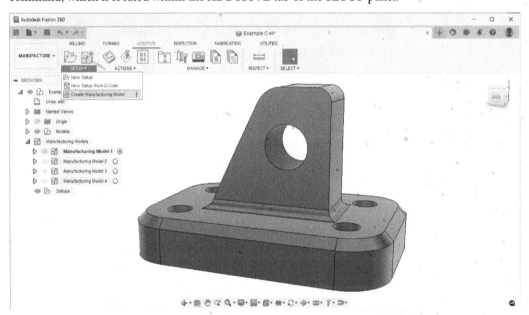

Figure 6.20 – Creating four manufacturing models based on the original design

Once we have created all four manufacturing models, we will activate **Manufacturing Model 1** by right-clicking it and choosing **EDIT**. In this first manufacturing model, we will demonstrate how to create two components that will be used for different manufacturing processes.

Our example model is already in the correct orientation for 3D printing. However, the large hole along the *Y*-axis cannot be 3D-printed accurately to our desired specifications without a support structure. In such cases, we have several options if we do not want to utilize support structures.

We can duplicate the original components and modify the hole so that it can be 3D-printed without support structures. Then, we can use the 3D-printed part as the stock material and create a hole using our original component as a reference to create the drilling operation in a milling setup and produce the finished design. Let's go through the steps of this process using Fusion 360 by duplicating our original component using the **Move/Copy** command located in the **SOLID** tab's **MODIFY** panel, as shown in *Figure 6.21*. After selecting the component we wish to duplicate, we will have to check the **Create Copy** checkbox and click **OK**:

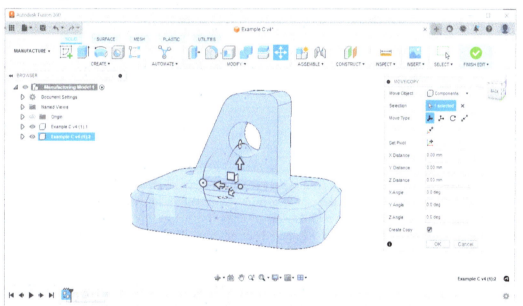

Figure 6.21 – Creating a duplicate component while editing Manufacturing Model 1

Once we have two components, we may want to rename our second component to better organize our design. In this example, I chose to name the second component Copy, as shown in *Figure 6.22*.

Next, we will have to create a sketch so that we can modify the hole. To start a sketch, we can choose the **CREATE SKETCH** command located in the **SOLID** tab's **CREATE** panel and select one of the faces along the *XZ* plane. Once we are in the sketching environment, we can project the profile of the hole and create a 45°-45°-90° right triangle, as shown in *Figure 6.21*. The reason why we are creating a 45-degree angle is that most 3D printers can manufacture parts without needing support structures up to 45-degree overhang angles. We will cover this topic in more detail in *Chapter 10*:

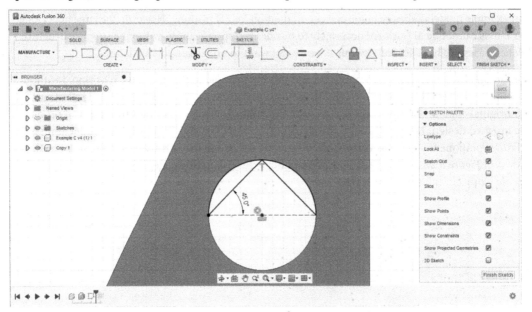

Figure 6.22 – Sketching a profile within the hole

Once we have created the desired profile, we can use the **EXTRUDE** feature located in the **SOLID** tab's **CREATE** panel. We will select the two profiles between the triangle and the top half of the circle. As shown in *Figure 6.23*, after extruding these profiles from one side of the hole to the other, we will end up with a component that can be 3D-printed without support structures. However, this new design has a hole that is shaped differently from the desired outcome. Therefore, after 3D-printing the Copy component, we will have to create a milling setup. Here, we must use the 3D-printed component as the stock and drill a hole using the geometry stored in the original component, named Example C:

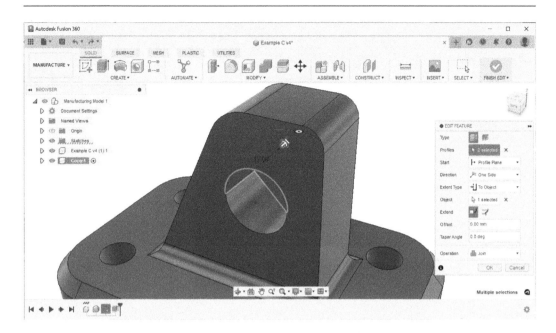

Figure 6.23 – Adding material to the hole to print without supports

Another alternative we have is to make a slightly larger hole so that we can print our component without the need for support structures or secondary milling operations. To demonstrate this workflow, let's activate **Manufacturing Model 2** by right-clicking it and choosing **EDIT**, as shown in *Figure 6.24*. We will create a new sketch on the same face we worked on in **Manufacturing Model 1**. In this case, we will also create a circle to match the projected geometry of the original hole and create two lines that are 45 degrees from the horizontal plane and are connected to the circle using a tangent constraint, as shown in *Figure 6.24*:

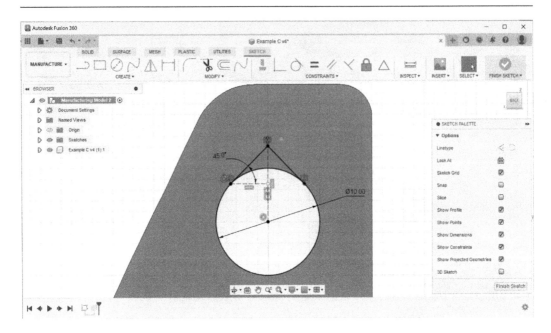

Figure 6.24 – Editing Manufacturing Model 2

After finishing the sketch, we can use the **EXTRUDE** feature and remove the new profile we have created from the original component. This action will create a larger hole that, when 3D-printed, can still accommodate a pin through the hole. However, unlike the version we had before making this modification, this part can now be 3D-printed without the need for support structures:

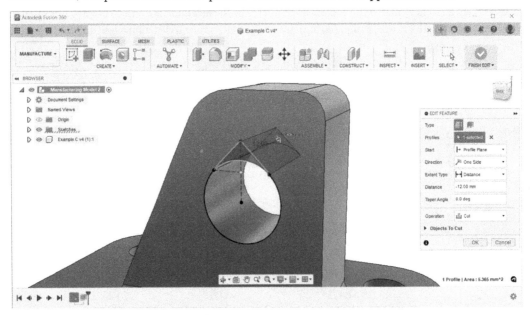

Figure 6.25 – Removing material from the hole to print it without supports

To take this example one step further, let's switch to **Manufacturing Model 3**, as shown in *Figure 6.26*. We will once again edit the manufacturing model and create a sketch on the same face we have worked on for the first two manufacturing models. In this manufacturing model, we will repeat the same steps we did in **Manufacturing Model 2** to create a circle and two lines that are 45 degrees away from the horizontal axis and our tangent to the circle. In addition, we will create a rectangle above the circle within the triangular profile:

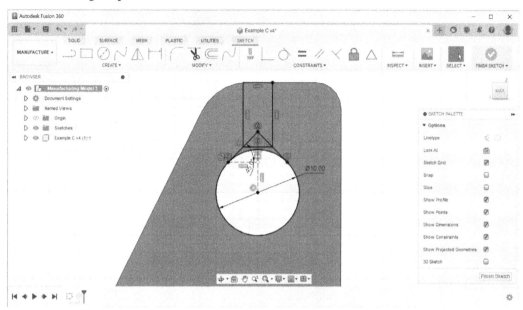

Figure 6.26 – Editing Manufacturing Model 3

Using these rectangular and triangular profiles, we can create another cut-out using the **EXTRUDE** feature, as shown in *Figure 6.27*. This new shape we created can not only be 3D-printed without the need for support structures but can also accommodate a pin that is slightly larger in diameter than the hole. This may be desirable as our 3D-printed geometry may not come out to our desired tolerances, or we may have a pin that is slightly bigger than our 3D-printed hole. The slot we cut out allows the 3D-printed part to slightly expand along the X-axis but will also provide a spring back to hold a pin in place:

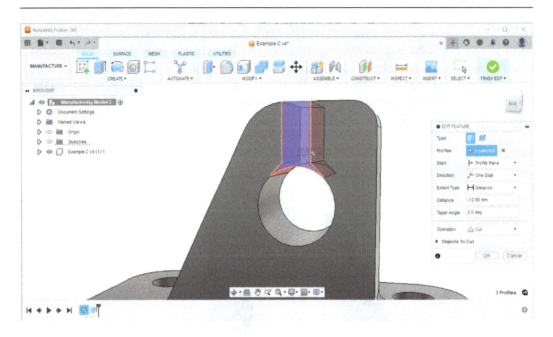

Figure 6.27 – Adding a slot for a tight fit and removing material from the hole

Next, we will demonstrate the fourth and final part modification. Unlike the first three manufacturing models, in **Manufacturing Model 4**, we will focus on the four small holes that are along the Z-axis. These holes can be printed without the need for support structures as their orientation is already aligned with the layer lines of our 3D printer.

Imagine a scenario where we want to add a slot feature to these holes so that they can hold a tight tolerance pin in place, similar to what we demonstrated with **Manufacturing Model 3**, but we do not wish to create a large cutout from our part to minimize the impact on its stiffness and aesthetics. In such cases, we can utilize additional sketching features within Fusion 360 to create unique customized slots for 3D printing. *Figure 6.28* shows our model from the top view, while we're editing **Manufacturing Model 4** with a sketch created on the *XY* plane. Using the projection of the hole, we can identify its center point and create a rectangular cutout zone, as well as a 90-degree **Center Point Arc Slot**:

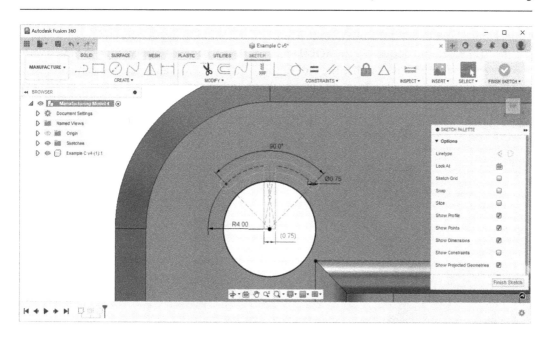

Figure 6.28 – Editing Manufacturing Model 4 to add an arc slot and a channel

Once we have the cut-out profile designed on one hole, we can utilize the **Mirror** command within the sketching environment and mirror our sketch lines along the X and Y axes, as shown in *Figure 6.29*. Using the **EXTRUDE** command located in the **SOLID** tab's **CREATE** panel, we can select the profiles we have sketched and remove them from the solid body:

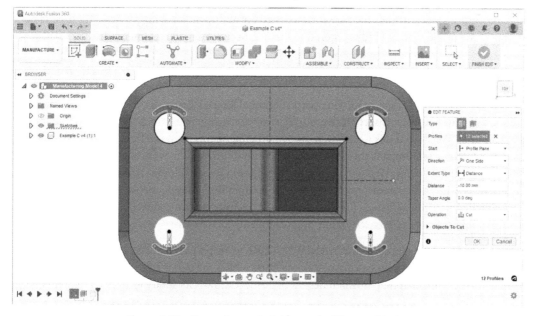

Figure 6.29 – Removing material from all of the small holes

As a final touch, we can select the two faces that are between our hole and the slot and create a fillet using the **FILLET** feature located in the **SOLID** tab's **MODIFY** panel. While creating the fillet, we can either enter a fillet radius or utilize the **Full Round Fillet** type. We can repeat the same process for all four holes and create fillets for eight faces, as shown in *Figure 6.30*:

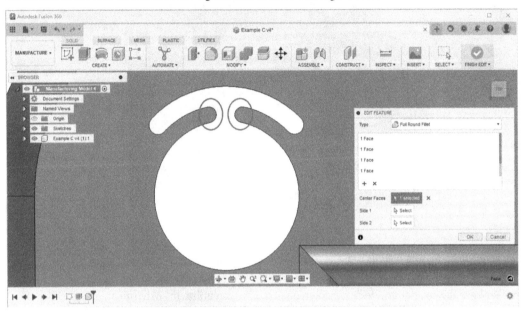

Figure 6.30 – Adding full round fillets to improve print bed adhesion

Depending on which combination of feature modifications we choose to utilize, we will end up with a manufacturing model that has the part edits we need for our manufacturing process. These edits do not impact the original design but are stored in their respective manufacturing models.

Summary

In this chapter, we talked about modifying parts for manufacturing purposes. We started this chapter by discussing how to use the **Insert Derive** workflow within Fusion 360 to manage our model changes for 3D printing. We utilized this workflow to aggregate components from multiple Fusion 360 designs and make changes that impact the downstream manufacturing processes but do not impact the upstream design document. We talked about the benefits of this workflow as it allowed us to bring in designs, along with their parameters, from multiple sources into a single manufacturing document. We also highlighted how utilizing this workflow blocks us from deleting source components, so long as those components are referenced in our manufacturing document.

Next, we introduced the concept of the **MANUFACTURE** workspace and how to create manufacturing models. We demonstrated how we can modify our parts within the manufacturing model of the same document. We highlighted that the main benefit of this workflow is its ability to store all relevant data in the same document that contains the Fusion 360 design. We also underlined the main drawback related to this workflow, which was the fact that we cannot easily aggregate components from other Fusion 360 designs.

We ended this chapter by focusing on making part modifications while editing multiple manufacturing models for the same component. We demonstrated four modifications we can make to a part to print common features, such as holes without the need for support structures while keeping tight tolerances between the hole and the pin that it will house.

Now that we are at the end of this chapter, you have all the knowledge you need to create manufacturing models, either within the same Fusion 360 design document or a new document to modify your parts for 3D printing. Using these techniques, you can edit your models for your design and assembly needs, as well as your manufacturing needs, and keep your changes for design and manufacturing separate but connected. In the next chapter, we will learn how to create our first additive setup.

7
Creating Your First Additive Setup

Welcome to *Chapter 7*. In previous chapters, we covered how to create and modify models for 3D printing. Now, we can start to go over how to set up a 3D print using Fusion 360. Before we can walk over to our 3D printer and press print, we need to define the process digitally on our slicer, which in this chapter will be Fusion 360. 3D printing using Fusion 360 can be as simple as pressing a single button if we first take the time to define mission-critical information, such as which printer we will be using and which material we will be printing with. In this chapter, we will dig a little deeper into how to use Fusion 360 effectively, so we can understand how the machine and print settings libraries work together to help us create an additive setup, so that we can go from a digital design to additive manufacturing.

In Fusion 360, a critical first step for 3D printing is to create an additive manufacturing setup. This simple yet powerful dialog includes several inputs for selecting our 3D printer, our desired material, and the print settings for that material. Understanding how to use Fusion's libraries and creating custom libraries with our 3D printers and print settings will benefit us greatly. In this chapter, we will also highlight various Fusion 360 preferences related to 3D printing, which will improve your experience and save you time.

In this chapter, we will cover the following topics:

- Fusion 360 preferences and settings for 3D printing
- Machine selection and creating an additive setup
- Print Settings and creating an additive setup

By the end of the first section, you will know how to locate and utilize various Fusion 360 preferences related to 3D printing. In the next section, you will learn where to locate your specific 3D printer digitally within the various machine libraries included with Fusion 360. By the end of the chapter, you will know how to copy, edit, and further manage your 3D printer and print settings from Fusion 360's libraries to your own library. In addition, you will have learned how to assign a printer and print settings to an additive setup. You will also be exposed to various print settings based on different 3D printing strategies and will learn how to edit them.

Technical requirements

All of the topics covered in this chapter are accessible to the personal, trial, commercial, startup, and educational Fusion 360 license types.

There are certain Fusion 360 additive machines, such as the **metal powder bed Fusion** (**MPBF**) 3D printers, that are not available until you gain access to the **Manufacturing Extension**. As with all other Fusion 360 extensions, the Manufacturing Extension is not available for personal use. As we will not cover topics specific to metal 3D printing in this chapter, you should be able to follow along with all the examples in this chapter with no issues.

If you have a commercial license for Fusion 360, you will also need access to the Manufacturing Extension (`https://www.autodesk.com/products/Fusion-360/additive-build-extension`) to create an additive setup with an MPBF machine.

Autodesk offers a 14-day trial for the Manufacturing Extension for all commercial users of Fusion 360. You can activate your trial within the extensions dialog of Fusion 360.

The lesson files for this chapter can be found here: `https://github.com/PacktPublishing/3D-Printing-with-Fusion-360`.

Fusion 360 preferences and settings for 3D printing

Fusion 360 is a feature-rich software for computer-aided design, simulation, and manufacturing. As such, it has numerous preferences and settings that can be customized to fit your specific workflow. In this book, we are mostly focusing on using Fusion 360 for additive manufacturing. We already touched on some of the basic preferences, such as how to set the default units and the ability to capture design history versus design using direct modeling, back in *Chapter 1*. In this chapter, we will dig a bit deeper and go over all the settings and preferences as well as some preview features that are available for improving our workflow for additive manufacturing.

We can start to simplify our use of Fusion 360 for additive manufacturing by accessing the **Preferences** dialog by selecting the **Avatar** icon in the top-right corner of the user interface as shown in *Figure 7.1*. The **Preferences** dialog has a browser on the left-hand side. By selecting the first entry, named **General**, we can access the preferences controlling the general user interface behavior. *Figure 7.1* shows how to set the default modeling orientation to **Z up**, which we covered in detail in *Chapter 1*:

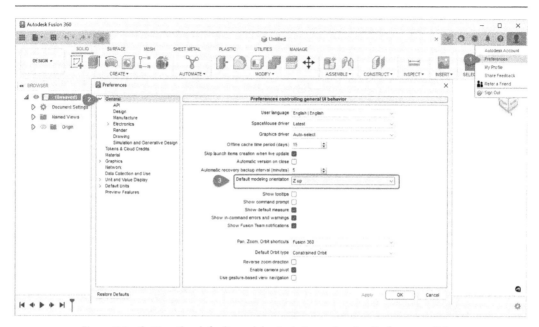

Figure 7.1 – Setting the default model orientation using the Preferences dialog

Under the **General** dropdown within the **Preferences** dialog, if we switch to the line item named **Manufacture**, we can access preferences that control the behavior of the user interface within the **Manufacture** workspace. As you may recall, we introduced the **Manufacture** workspace in *Chapter 6* and used it to make model modifications using manufacturing models. *Figure 7.2* shows how to use **Enable Cloud Libraries** to save and select things such as our 3D printers and print settings to the active Fusion Team. We can also set the user interface such that every time we access the **Manufacture** workspace, we will automatically be taken to the **ADDITIVE** tab. If we scroll down within the preferences that control the behavior in the **Manufacture** workspace, we can also see settings specific to additive manufacturing, such as options that can automatically generate support, automatic orientations, and additive toolpaths upon the creation of the function or an additive setup. We will not be changing the default settings for those three options as the default preferences work well for additive manufacturing. In the future, as you customize Fusion 360, you may need to refer back to those settings and modify them as needed. For example, if you always print parts without needing support structures, you can activate **Generate Additive Toolpath on Setup creation** and Fusion 360 will automatically create the slices and the toolpath without waiting for your explicit command:

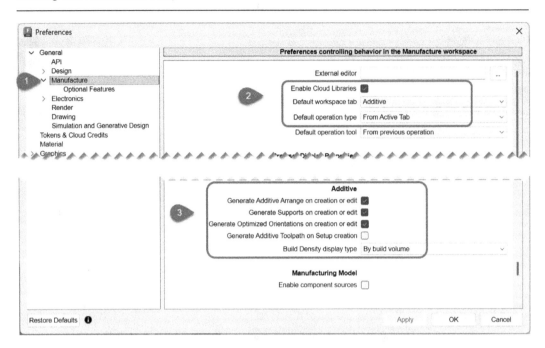

Figure 7.2 – Fusion 360 preferences specific to the Manufacture workspace

Another important preference I would like to highlight can be found under the **API** section of the **Preferences** dialog. In the preferences for scripting and programming, Fusion 360 shows **Default Path for Scripts and Add-Ins** as shown in *Figure 7.3*. We will not be changing this path in this chapter. However, seeing this path gives us a clear indication of where Fusion 360 saves custom assets, such as machine files and print setting files and posts. As highlighted in *Figure 7.3*, any machine or print setting file you save into your local folder automatically gets stored in your user profile within the `AppData\Roaming\Autodesk\` folder within the appropriate subfolder, such as `machines` or `PrintSettings`.

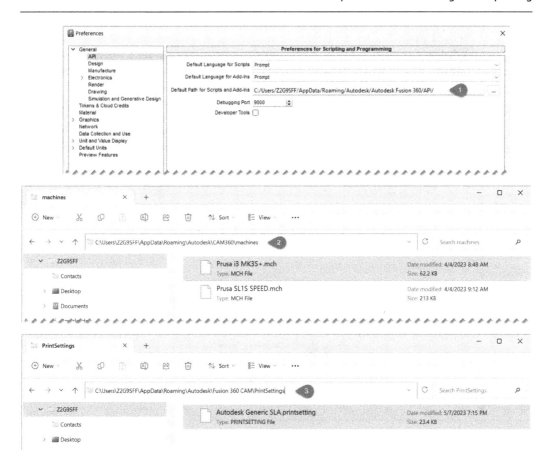

Figure 7.3 – Default Path for Scripts and Add-Ins indicates where
Fusion 360 saves machine and print settings files

The final preference I would like to highlight is accessible within the **Preview Features** section of the **Preferences** dialog. As shown in *Figure 7.4*, by using this option, we can see a **Preferences to try Preview Functionality** dialog. If we set our filter to **Manufacture**, we can turn on several pre-release functionalities that Fusion 360 offers its users before they become a part of a regular release.

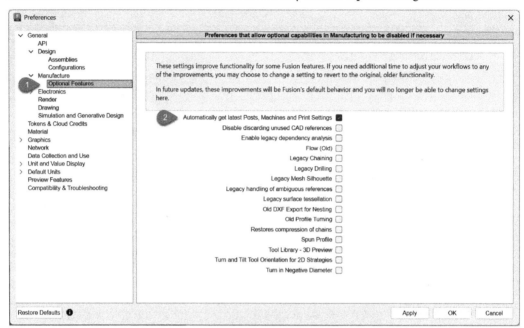

Figure 7.4 – Manufacturing-related preview features within the Preferences dialog

Once we have all the preferences set, as shown in the previous figures, we can select **OK** and proceed to modify our settings for the **Manufacture** workspace. In order to make changes to the **Manufacture** workspace, we can simply open a new Fusion 360 design document and switch to the **Manufacture** workspace using the workspace switcher located in the top-left portion of Fusion 360's user interface. When we are in the **Manufacture** workspace, using the navigation bar located in the bottom-center portion of the canvas, we can change settings, such as the display effects, and turn off the ground plane, as shown in *Figure 7.5*. The reason why we may want to turn off the ground plane is that sometimes the ground plane may be above or below the build platform of our 3D printer and could create a crowded display, blocking our view while we are working on our additive setup.

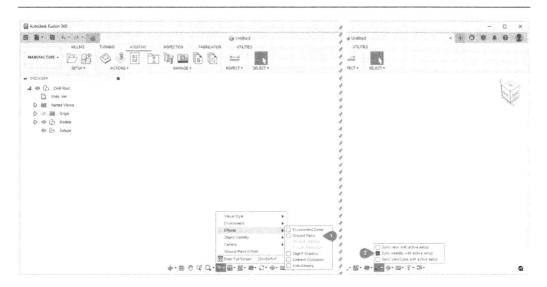

Figure 7.5 – Display- and visibility-related settings for the Manufacture workspace

Another setting I suggest changing while working on additive setups is the synchronization of the visibility. I recommend setting **Sync visibility with active setup** using the navigation bar at the bottom of Fusion 360's user interface within the **Manufacture** workspace as shown in *Figure 7.5*. When this setting is active, Fusion 360 will automatically hide components that do not participate in the active additive setup.

In this section, we covered various settings and preferences that will help us while creating and working with additive setups within the **Manufacture** workspace of Fusion 360. In the next section, we will focus on creating an additive setup and selecting a 3D printer for that setup.

Machine selection and creating an additive setup

Now that we have customized Fusion 360 for additive manufacturing, it is time for us to use the **Manufacture** workspace and create our first additive setup. To create an additive setup, the first thing we need to do is to select our 3D printer. In this section, we will focus on choosing a 3D printer from Fusion 360's machine library. Fusion 360 hosts its machine libraries for additive, subtractive, and inspection machines online at `https://cam.autodesk.com/machineslist`. If our 3D printer does not exist in Fusion 360's online library, we will have to create it manually. We will not go over creating a custom 3D printer in this section.

Figure 7.6 shows the online machine library for Fusion 360, as displayed on a web browser. This web page is equipped with a search field, which allows us to type the name, or a portion of the name, of the machine we are searching for. The web page also has two drop-down fields to help us filter the results. The first filter allows us to screen the results based on technology, such as additive, milling, turning, and so on. The second filter, which is shown in *Figure 7.6* in its expanded state, allows us to screen the results based on the manufacturer. Once you find the specific 3D printer you are interested in using with Fusion 360, you can press the **Download** button, and you will have access to the `*.MCH` file on your computer's `Downloads` folder:

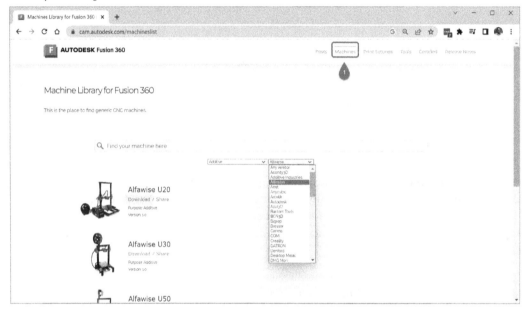

Figure 7.6 – Filtering Fusion 360's additive machines based on the manufacturer

Another way you can access the machine library is by creating a manufacturing setup. To create an additive manufacturing setup, you can open any Fusion 360 design document and switch to the **Manufacture** workspace. Since we set our preferences to default to the **ADDITIVE** tab when using the **Manufacture** workspace in the previous section, we will automatically be in the **ADDITIVE** tab. After selecting the **New Setup** command located within the **SETUP** panel, Fusion 360 will automatically create a manufacturing model based on the original design and will display the **SETUP** dialog as shown in *Figure 7.7*. The **SETUP** dialog is made up of two tabs: **Setup** and **Post Process**. Within the **Setup** tab, the first thing we must do is select our machine for the additive operation we are about to create:

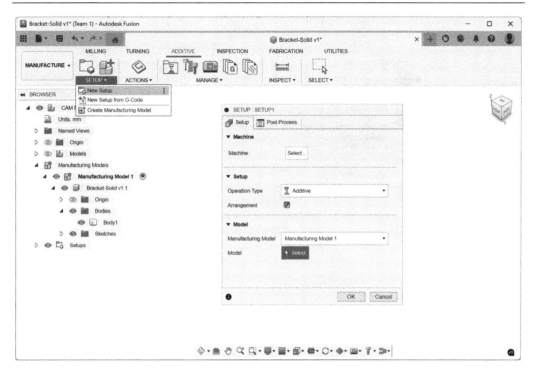

Figure 7.7 – Using the SETUP dialog to create an additive setup

After pressing the **Select…** button for the machine selection in the **SETUP** dialog, Fusion 360's machine library will be displayed. The machine library is made up of three columns. The first column on the left is the browser for various locations we can search for our machine. The middle column shows the various machines we can select. The right-most column allows us to filter our machines based on capabilities and technologies and displays information about the selected machine. The **Machine Library** dialog also has a search bar on the top left. we can see all the machines the Fusion 360 machine library hosts on the web within the **Machine Library** dialog, as shown in *Figure 7.8*. If we expand the list named **Fusion 360 library**, we will see all the machine vendors that **Fusion 360 library** contains a machine for. In *Figure 7.8*, we can also see the five different **Fused Filament Fabrication** (FFF) printers from Alphawise within the Fusion 360 library:

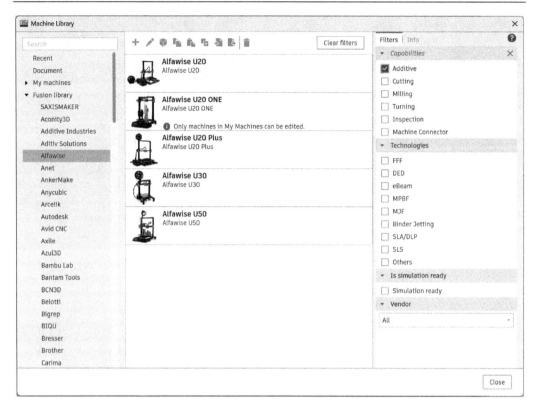

Figure 7.8 – Fusion 360's library of machines filtered for Additive

Using the search bar in the top-left corner of the machine library, we can also search for a specific machine. In *Figure 7.9*, after typing MK3 into the search field, we see the three FFF printers from the machine vendor Prusa Research. We can select the specific printer we wish to use for our additive setup, press **Select** in the bottom-right corner of the **Machine Library** dialog, and proceed to the **Print Settings** selection.

Alternatively, as shown in *Figure 7.9*, after selecting our printer, we can right-click and copy it, so that we can paste it into one of many locations within **My machines**, such as **Cloud**, **Local**, and **Linked**.

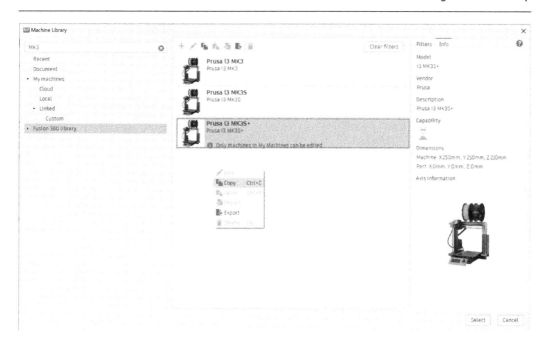

Figure 7.9 – Searching for a specific printer and copying within Machine Library

Please keep in mind that the reason why we see **Cloud** as an option under the **My machines** section within the browser is that, in the previous subsection, we enabled **Cloud** libraries as was shown in *Figure 7.2*. In *Figure 7.3*, we also highlighted where our local machines were being saved on our computer. But, until now, we have not covered **Linked** libraries. So, let's take a closer look at **Linked** libraries. As shown in *Figure 7.10*, we can utilize a **Linked** library by selecting it, right-clicking on it, and choosing a folder to link. This will allow us to select a specific folder on our computer that will store all the files associated with that library:

Figure 7.10 – Linking a local folder to save machine files

Once we have a folder that is linked, all the machines that were present in that folder will be displayed in the **Machine Library** dialog. If we do not have any machines in our **Linked** folder, we can simply copy and paste machines from other locations, such as **Fusion 360 library**, as shown in *Figure 7.11*:

Figure 7.11 – Pasting a previously copied machine into a Linked folder

After we paste our selected machine within the **Linked** folder, we can also check to make sure that our selected machine is located in our **Linked** folder. We can do that by using Windows Explorer or Mac Finder and navigating to our selected folder. *Figure 7.12* shows that after copying and pasting the Prusa i3 MK3S+ machine from **Fusion 360 library** to our **Linked** folder, it is displayed within the machine library under the **Linked** custom folder, and the same folder in Windows Explorer now contains a new file called Prusa i3 MK3S+.mch:

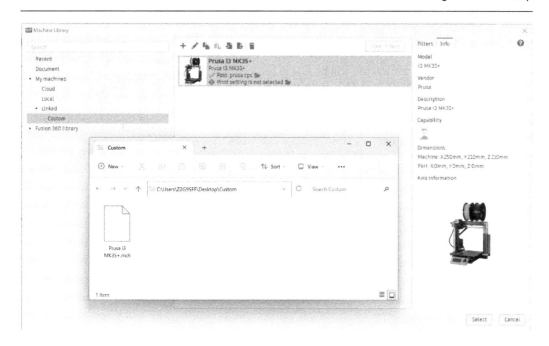

Figure 7.12 – The Linked folder is updated after pasting a new machine

If we were to paste the previously copied machine over to the folder named **Cloud** under the **My machines** section, those machines would get saved within the assets section of our active Fusion Team. As shown in *Figure 7.13*, we can locate the *.mch file by going to the **Assets** folder and then navigating to the folder named **CAMMachines**. Within this location, we will be able to find the Prusa i3 MK3S+.mch file:

Figure 7.13 – Locating the cloud library of machines within the Assets folder

If you have access to Fusion 360's **Home** tab experience, in addition to locating the **Assets** folder in the **Data** panel, you will be able to see it in the **Home** tab as shown in *Figure 7.14*. After selecting the **Home** icon in the top-left corner, you will be able to see the projects that are accessible to your active team. The **Assets** folder will be available within **Projects** and the **CAMMachines** folder will store the machines you have saved within your **Cloud** library:

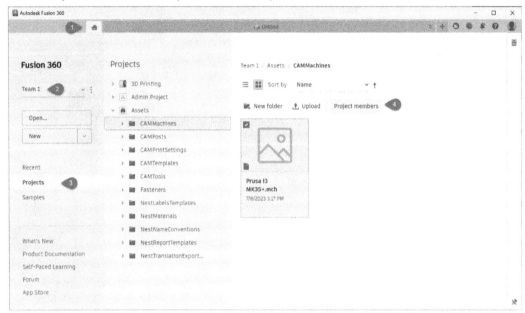

Figure 7.14 – Locating the Cloud library of machines within the Home tab

The main difference between the **Local** and **Linked** libraries versus the **Cloud** library relates to collaboration. Library items, such as machines, print settings, tools, and posts you save to your **Cloud** library, are all accessible to every project member within your active team. This means if you have five users within your team, they will all have read/write access to all the assets you saved to your **Cloud** libraries. *Figure 7.14* also shows a full list of all possible assets, ranging from various libraries to templates and materials.

In this section, we covered selecting a machine during an additive setup and managing our machine libraries. In the next section, we will talk about how to do the same for print settings and will look at the different print settings associated with different additive technologies.

Print settings and creating an additive setup

We can utilize the **Print Settings** library to find and download the latest print settings available from Autodesk for all types of 3D printers, just as we used the web-based **Machine Library** in the previous section. Autodesk hosts the **Print Settings** library for Fusion 360 at `https://cam.autodesk.com/printsettingslist`. After launching the web page, we can use the search bar to find the print settings we are looking for or use the filters for 3D printing technology and machine vendor to narrow down the extensive list of print settings available on this web page. Once we find the print setting we need, as shown in *Figure 7.15*, we can simply download it to our computer using the **Download** button:

Figure 7.15 – Fusion 360's Print Settings library on the web

We can also access the entire **Print Settings** library of Fusion 360 during an additive setup creation. After selecting our machine, as we did in the previous section, we can press the **Print Settings** button within the **SETUP** dialog and choose from the available options within the Fusion 360 library of print settings. Similar to how we copied and pasted machines from the Fusion 360 library to our **Linked** folders, we can copy and paste print settings from the Fusion 360 **Print Settings** web page or the Fusion 360 library. After downloading a print setting from the web, we can import it to our **Local**, **Cloud**, or **Linked** print setting libraries and select that print setting from our preferred library during the additive setup as shown in *Figure 7.16*:

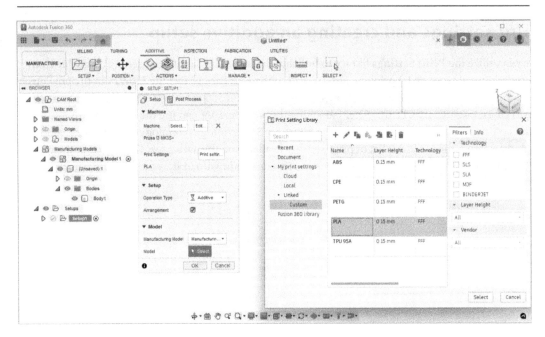

Figure 7.16 – Selecting a print setting from Fusion 360's Print Setting Library

Once we have a print setting saved within a folder under **My print settings**, such as the **Linked | Custom** folder shown in *Figure 7.17*, we can modify that print setting by selecting it, right-clicking it, and editing it. Editing is only available after a machine or a print setting is saved under the **My machines** or **My print settings** subsection of **Fusion 360 library**:

Figure 7.17 – Editing a print setting located under My print settings

Print Setting Editor is displayed upon pressing the **Edit** command for a custom print setting, which will allow us to customize all the information based on our preferences. *Figure 7.18* shows a typical print setting for an FFF print setting. This print setting happens to be the default print setting for printing on a single extruder FFF printer using the PLA material. **Print Setting Editor** is made up of two columns. On the left-hand side, we can see headings such as **Information**, **General**, and **Body Presets**. Under **Body Presets**, we have two headings by default: **Normal** and **Strong**. **Normal** happens to be the default setting for **Body Presets**, as denoted by the gear icon with a checkmark to the left of the text. We will cover interacting with **Body Presets** and assigning print settings per component in more detail in *Chapter 8*:

Figure 7.18 – Customizing the PLA print setting for an FFF machine

Under the **Body Presets** section, within the **Normal** body preset, there are multiple subsections, such as **Extrusion**, **Shell**, and **Infill**. *Figure 7.19* shows the input fields for the **Infill** subsection of the **Normal** body preset. Using this entry, we can customize the **Infill** pattern from the default option of **Gyroid** to a different **Infill** pattern. Or we can change **Infill Density** from 25% to 100% to make it a fully dense print if desired:

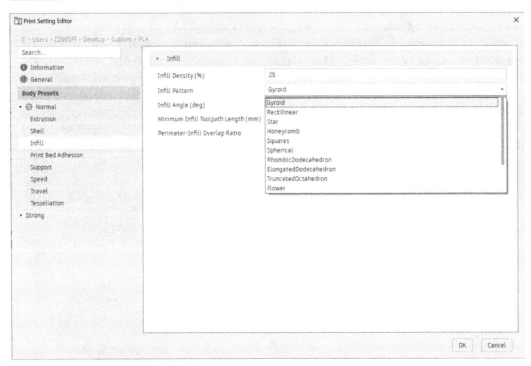

Figure 7.19 – Modifying the Infill inputs for an FFF print setting

Another common **Body Presets** setting to modify is **Print Bed Adhesion**. *Figure 7.20* shows **Print Bed Adhesion** for the **Normal** body preset. As you can see in the figure, **Print Bed Adhesion** has been enabled and **Print Bed Adhesion type** has been set to **Skirt** by default. You can change **Print Bed Adhesion type** to **Brim** or **Raft** and modify the associated inputs based on the adhesion type you have selected within this dialog:

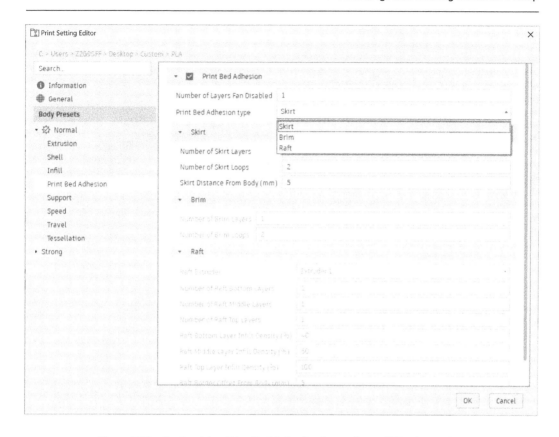

Figure 7.20 – Customizing Print Bed Adhesion inputs for an FFF print setting

Print Setting Editor also allows us to search a given input field by its name using the search bar located in the top-left corner of the dialog. *Figure 7.21* shows the outcome of searching for the phrase `number of perimeters`. This entry exists twice for this print setting, as shown in the figure. The first entry is for the **Normal** body preset and the second entry is for the **Strong** body preset. After selecting **Number of Perimeters (Normal)** from the list of available options, the user interface automatically switches to the **Shell** selection of the **Normal** body preset and highlights the **Number of Perimeters** entry field with a blue line beneath it. If we hover over the input field, we are also able to see a tooltip for this entry, which reads **Number of perimeter beads around each layer**:

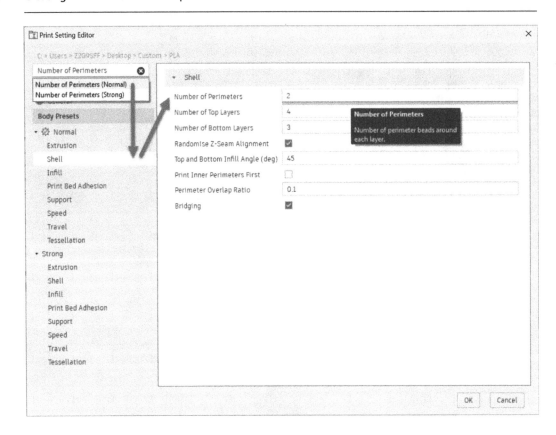

Figure 7.21 – Using the search bar for locating and adding input to modify within Print Setting Editor

So far, we have mainly focused on print settings and machines for FFF. However, Fusion 360 supports the creation of additive setups for more than FFF printers. If we were to select a 3D printer that utilizes **stereolithography (SLA)** while we were creating our additive setup, then after pressing the **Print Settings** button, we would be able to select a print setting that was appropriate for the type of printer we had just selected. *Figure 7.22* shows what **Print Setting Editor** looks like for the print setting named **Prusa Orange Tough - 0.05mm Normal**, which is a resin that can be printed on a Prusa SL1 printer, which is an SLA 3D printer. As you can see, the list of available options for this print setting is very different from the PLA print setting we have been editing for an FFF printer. There are similarities between the two, such as inputs for layer thickness. However, for resin-based printers, we have access to editing additional information, such as exposure time, which can be controlled within **Print Setting Editor**:

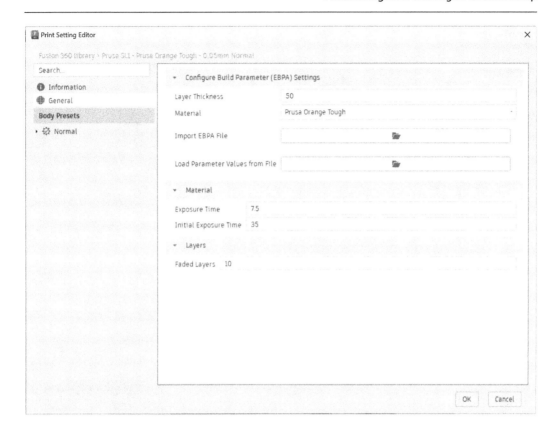

Figure 7.22 – Print Setting Editor for a Prusa Orange Tough resin to be printed on a Prusa SL1 printer

The print settings for all SLA printers are not the same, and they can look different based on the specification of the machine manufacturer and the slice file required by the machine. *Figure 7.23* shows a good example of a generic SLA printer that utilizes a ZIP file full of SLC (slice) files. You can access this print setting after choosing a generic SLA printer from the machine library. When editing this print setting, you can customize the layer thickness and have control over whether you want to create unique files per part and for supports and how you wish to name each file:

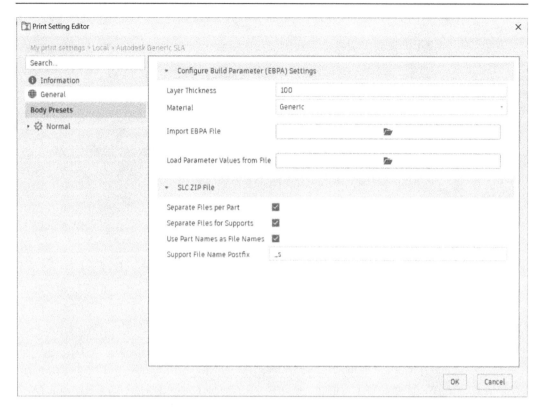

Figure 7.23 – Print Setting Editor for a generic SLA printer

Even though both the DLP and SLA printers utilize resin as the raw material to 3D print with, they use different technologies to print that resin, and the print settings for such printers also reflect that difference in technology. After selecting an Autodesk Generic DLP printer from the machine library, you will be able to select the Autodesk Generic DLP as the print setting. If you edit that print setting as shown in *Figure 7.24*, you will see yet another unique list of print settings you can modify. Similar to the FFF and SLA print settings, you can control the layer thickness for a DLP print setting. However, unlike the other technologies, DLP printers generally utilize image files to print each layer. In this print setting, you can customize the image size, the image quality, and the naming associated with each image file (*.png) during the export process. Before slicing your model and exporting the image files, you will want to make sure you edit the **Size X** and **Size Y** fields of this print setting to match the printable build area dimensions of your specific DLP printer:

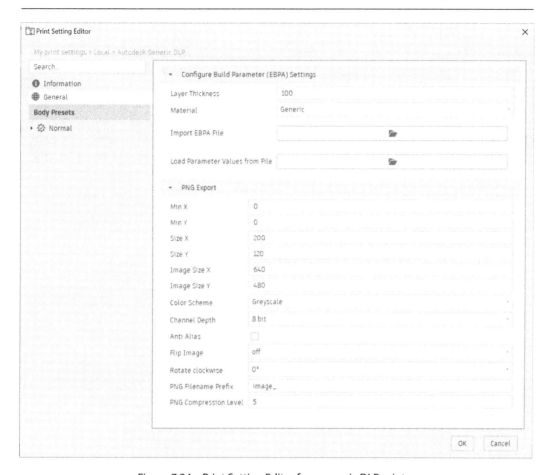

Figure 7.24 – Print Setting Editor for a generic DLP printer

Even though we are not focusing on 3D printers using MPBF technology in this chapter, I also want to take the opportunity to showcase what a print setting looks like for this technology. As mentioned earlier in the chapter, you cannot select an MPBF 3D printer as a part of your additive setup unless you have access to the Manufacturing Extension. However, if you want to follow along and duplicate the workflow shown in *Figure 7.25* to better understand the print settings associated with MPBF, you can simply find an MPBF print setting from Fusion 360's online library and download it. Once you have a local copy of an MPBF print setting, you can import it into one of your **Local** or **Cloud** libraries, as we covered in the previous section. You don't need to do this during the creation of an additive setup. You can access all manufacturing-related libraries within the **Manufacture** workspace's **MANAGE** panel within any tab.

Figure 7.25 shows **Print Setting Editor** when editing the `Autodesk Generic MPBF 100 micron` print setting. This print setting, much like all the other print settings we have demonstrated up until now, allows us to modify the layer thickness. In addition, we can designate our metal powder in the **Material** dropdown, which has an impact downstream when we wish to use this print setting for a process simulation, which will cover in more detail in *Chapter 13*. We also have additional controls over the path the laser will follow while fusing the powder for different parts of the model, such as its **Upskin**, **Downskin**, and **Supports**. The Autodesk Generic MPBF 3D printer, along with this print setting, allows for the creation of multiple slice file types, such as `SLC` and `Gcode`. Therefore, the print setting also incorporates various options that control how those files are created and named:

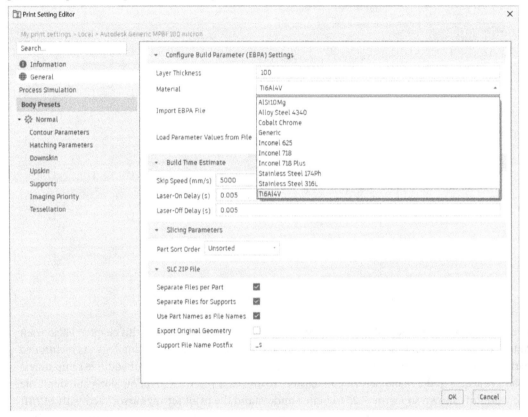

Figure 7.25 – Print Setting Editor for a generic MPBF printer

Now that we have covered a variety of print settings for different additive technologies, let's put everything we have learned into practice by creating our first additive setup. To demonstrate this, we will be using a model we have created in *Chapter 6* named **Spur Gear**. We are purposefully using this geometry as it is already oriented such that it has a flat surface that touches the X-Y plane.

After opening the design in Fusion 360, we can switch to the **Manufacture** workspace using the workspace switcher. Within the **ADDITIVE** tab, we can click on **New Setup**, which is in the **SETUP** panel. Within the **Setup** tab of the **SETUP** dialog, click on the **Select…** command next to the **Machine** input, and select **Prusa i3 MK3S+** as our machine from **Fusion 360 library**. Using the **Print Setting** selector, let's select **PLA** as our **Print Setting**. Using the model selector at the bottom of the dialog, we can double-click on **Spur Gear** in the canvas or select the **Spur Gear** component from the browser within **Manufacturing Model 1**, as shown in *Figure 7.26*.

After having completed all the selections related to the machine, print settings, and model, we have completed our input for this additive manufacturing setup. One other critical piece of information we have not highlighted so far is arranging parts within the build volume. By default, a **SETUP** dialog for an additive operation automatically has the **Arrangement** checkbox checked:

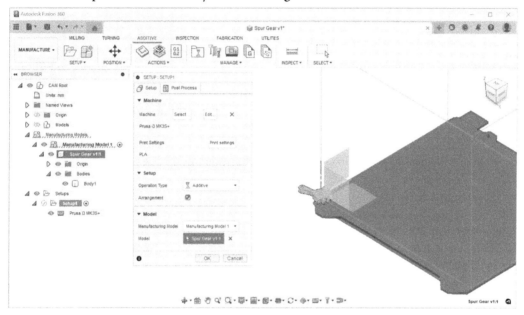

Figure 7.26 – Additive setup inputs completed for a single component

After selecting **OK** in the **SETUP** dialog, Fusion 360 will automatically display the build volume of the printer we have selected within the canvas and create **Setup 1**. The ribbon user interface will then change slightly, and we will see new icons appear in the ribbon under the **ADDITIVE** tab. If you compare the ribbon in *Figure 7.27* to the previous figure, you will see some new icons under the **POSITION** panel and a new panel called **SUPPORTS** appears. Once the additive setup is created, Fusion 360 will display a new dialog called **ADDITIVE ARRANGE**. We will focus on arranging our parts within the build volume in the next chapter in detail. But for this example, let us go ahead and press **OK** in the **ADDITIVE ARRANGE** dialog, which will automatically arrange this component within the build area of this printer:

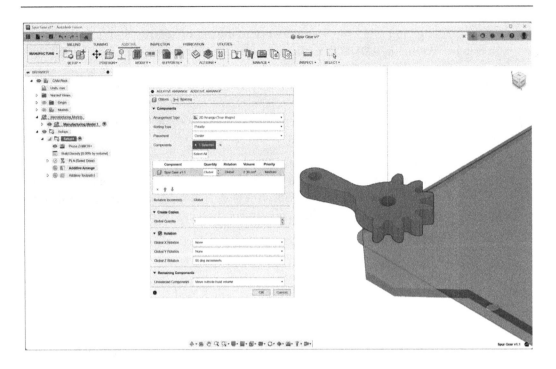

Figure 7.27 – Utilizing ADDITIVE ARRANGE within an additive setup

After the arrangement is complete, our part will be touching the build plate as the platform clearance input was 0 during the arrangement. An **ADDITIVE ARRANGE** operation automatically places all selected components around the center of the build platform. After pressing **OK** within the **ADDITIVE ARRANGE** dialog, Fusion 360 will perform the necessary calculations and move the part to the center of the build platform:

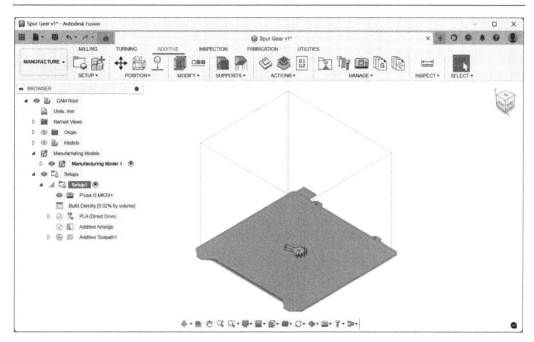

Figure 7.28 – Our first additive setup is now complete

Performing **ADDITIVE ARRANGE** also adds a new line item in the browser called **ADDITIVE ARRANGE**, which can be edited and or regenerated in the future if needed. Using this function, we can arrange one or more components within the build area of a given printer. As this is an FFF printer, the arrangement type was set to arrange components in 2D by default. If we were to create an additive setup using **selective laser sintering** (**SLS**), a **Multi Jet Fusion** (**MJF**) printer, or a binder jetting printer, we could have also used the 3D arrangement type and automatically organized our parts within the build volume of our selected 3D printer.

With that final step, we are now done with our first Fusion 360 additive setup using an FFF printer.

Summary

In this chapter, we have focused mainly on Fusion 360's settings and preferences and how to customize them in order to make our additive manufacturing workflow seamless. We have made changes to settings that control Fusion 360's display behavior and functionality related to automatically switching to the **ADDITIVE** tab when we enter the **Manufacture** workspace.

We have also enabled certain review features to utilize functionality that is considered pre-release. After customizing our settings and preferences, we learned about manufacturing-related libraries, such as the machine library and the **Print Settings** library. We explored the online libraries for 3D printers and Print Settings Autodesk hosts on the web. We demonstrated how to download machines, print settings from the web, and utilize them within the Fusion 360 user interface by saving them locally, as linked libraries, and as cloud-saved libraries, which are accessible to all our team members.

We highlighted the similarities and differences in print settings for different 3D printing technologies, such as FFF, SLA, DLP, and MPBF. We concluded the chapter by putting all our newfound knowledge into action by creating our first additive setup for a single component. Using everything we have learned in this chapter, you should feel confident when you are creating manufacturing setups, choosing the component you want to print, and generating your additive setups in Fusion 360.

In the next chapter, we will take the concept of part arrangement one step further and will focus on arranging and orienting multiple components within our 3D printer's build volume.

Part 3:
Print Preparation – Positioning Parts, Generating Supports, and Toolpaths

In this part, we will introduce the various part positioning and arranging options within a given Fusion 360 additive setup. We will highlight the various additive manufacturing technologies and the requirements around part arrangement and orientation. Then, we will introduce how to select a print setting for the additive manufacturing technology we will use in our setup. We will learn how to customize print settings and how to apply print settings to individual sections of our models. In the subsequent chapters, we will learn how to create support structures for additive technologies, such as fused filament fabrication and stereolithography. We will end this part by introducing how to slice our models and simulating the additive tool path, layer by layer.

This part has the following chapters:

- *Chapter 8, Arranging and Orienting Components*
- *Chapter 9, Print Settings*
- *Chapter 10, Support Structures*
- *Chapter 11, Slicing Models and Simulating the Toolpath*

8

Arranging and Orienting Components

Welcome to *Chapter 8*. In previous chapters, we highlighted how to design from the ground up as well as how to modify existing designs for 3D printing. At the end of *Chapter 7*, after having selected a printer and a print setting, we created our first Additive Manufacturing setup in order to 3D print a single component. In this chapter, we will build on that knowledge and create additive setups for different manufacturing technologies with one or more components and explore the different tools available within Fusion 360 to position those components.

Whether you're printing a single component or hundreds of them at the same time, arranging them within the build volume and orienting them based on the desired outcome is an important consideration to get a successful print. This chapter will go over the various 3D-printing technologies and how to best arrange and orient components for each technology. In this chapter, we will highlight how to manually translate and orient components. We will also showcase how to automatically orient parts and arrange them based on a given criteria such as minimizing build height or finding an orientation that will result in a minimum support-structure volume.

In this chapter, we will cover the following topics:

- Positioning components manually
- Orienting components automatically
- Arranging components automatically

By the end of the first section, you will have learned how to manually position components within the build volume of your printer. In this section, you will also learn various techniques to effectively translate and rotate your parts so that they are above the build plate of your printer and lying flat on a surface of your choosing. At the end of the second section, you will have learned how to conduct an orientation study for a given part and how to choose an orientation to print your parts based on the 3D-printing technology and your printer's features and limitations. By the end of the chapter, you will have learned how to automatically arrange your components within the 3D build volume of your printer. In addition, you will also be able to inspect your setup and perform basic statistics to find the arrangement density of your parts within your printer's build volume.

Technical requirements

All topics covered in this chapter are accessible to personal, trial, commercial, startup, and educational Fusion 360 license types.

There are certain Fusion 360 additive machines, such as the **Metal Powderbed Fusion** (**MPBF**) 3D printers, that are not available until access is gained to the Manufacturing extension. As with all other Fusion 360 extensions, the Manufacturing extension is not available for use with personal licenses. As we will not cover topics specific to metal 3D printing in this chapter, you should be able to follow along with all the examples in this chapter with no issues.

If you have a commercial license of Fusion 360, to create an additive setup with an MPBF machine, you will also need access to the Manufacturing extension (`https://www.autodesk.com/products/Fusion-360/manufacturing-extension`).

Autodesk offers a 14-day trial for the Manufacturing extension for all commercial users of Fusion 360. You can activate your trial within the extensions dialog of Fusion 360.

The lesson files for this chapter can be found here: `https://github.com/PacktPublishing/3D-Printing-with-Fusion-360`

Positioning Components Manually

When preparing our part for 3D printing, one of the primary functions we use in our slicer software is positioning components. Fusion 360 has multiple tools we can use for translating and orienting our parts within the build volume of our 3D printer. The first method we highlight in this section is the use of the **Move Components** command, located within the **POSITION** panel of the **ADDITIVE** tab within the **MANUFACTURE** workspace. In order to demonstrate how to use this functionality, we will create a new additive setup using a `Prusa i3 MK3S+` 3D printer with a `PLA` print setting.

Our first additive setup in this chapter will utilize two components named Component 6 and Component 15 within an assembly called Spark Max Motor Controller. We will create this additive manufacturing setup using the **NEW SETUP** command located in the **MANUFACTURE** workspace in the **ADDITIVE** tab's **SETUP** panel, as demonstrated in the previous chapter. During the setup creation, we will uncheck the **Arrangement** checkbox, which will allow us to have full control over the position of each component. We will discuss how to use the automatic arrangement functionality this checkbox initiates when activated in the final section of this chapter.

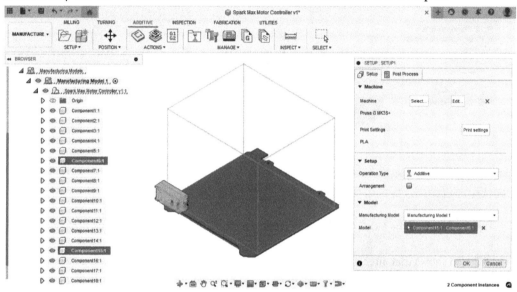

Figure 8.1 – Creating an additive setup with two components

While creating the additive setup, Fusion 360 automatically turns off the visibility of components that do not participate in the additive setup and displays the build plate and the build volume of the printer we have selected, along with the two components that will take part in this additive setup. As you can see in *Figure 8.1*, the two components do not move into the build volume automatically, as we unchecked the arrangement checkbox during the setup creation. Instead, they stay in place, which means we have to move them into the build volume manually.

When positioning parts within the additive setup, we have several options. The first option is to utilize the **MOVE COMPONENTS** command located within the **ADDITIVE** tab's **POSITION** panel. Once activated, the **MOVE COMPONENTS** dialogue allows us to select one or more components and translate and/or rotate them within the active setup. This action moves the components and captures their new position within the Manufacturing Model that is being used in the active setup. *Figure 8.2* shows how the selected component is translated on the *X*, *Y*, and *Z* axes by distances of 100 millimeters, 100 millimeters, and 60 millimeters respectively.

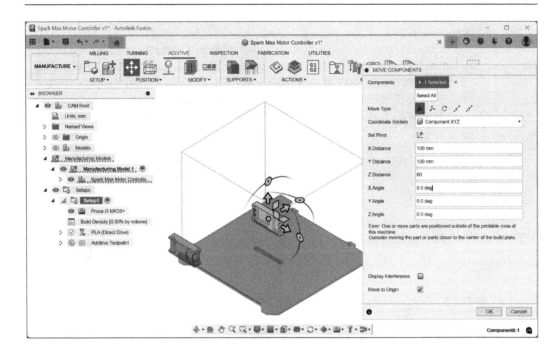

Figure 8.2 – Using the MOVE COMPONENTS dialog to position a part within the build volume

Figure 8.2 also shows that after moving the first component, we still have the second component, which is not fully within the build volume of our printer. Such components are displayed with a red highlight around their edges, indicating that those parts are positioned outside of the printable zone of the machine. Fusion 360 also displays an error message within the move components dialog indicating such issues.

If at this point, we wish to move a different component, we can select it, and then unselect the first component by left-mouse-clicking on the components in the canvas. Next, we can enter the distance and the angle by which we wish to move the second component and click **OK**.

Another option we have when positioning components is to move them within the canvas by selecting the center of the component with our mouse and dragging them on the screen as shown in *Figure 8.3* while the **MOVE COMPONENTS** dialog is active. This action is akin to performing a **MOVE/COPY** action with **Move type** of **Free Move** for components that are not grounded within the **DESIGN** workspace.

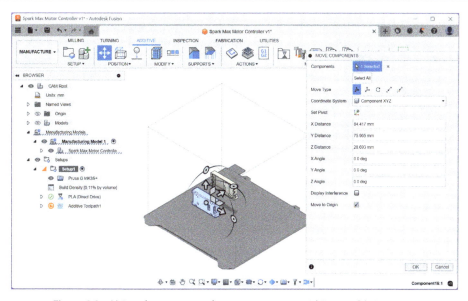

Figure 8.3 – Using the mouse to drag a component within an additive setup

As you can also see in *Figure 8.3*, the increment Fusion 360 uses during this free part motion has three significant figures for any translation or rotation. If we wanted to change the precision with which we translate or rotate components using the drag and drop technique, we would need to go back to our preferences dialog and change the general precision from the default, which is **0.123,** to either a less precise or more precise option. As a reminder, **General precision** can be found within the **Unit and Value Display** section of the **Preferences** dialog, as can be seen in *Figure 8.4*.

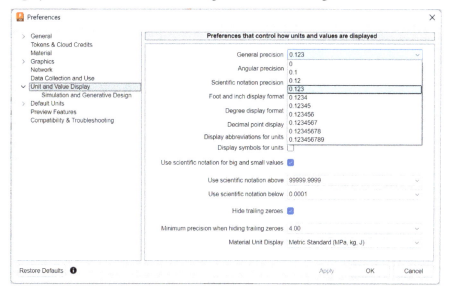

Figure 8.4 – General precision can be edited within Fusion 360's Preferences dialog

The **MOVE COMPONENTS** dialog has additional functionality, which allows us to rotate our components as well. After selecting a component to rotate, we can type in an angle within the **X Angle**, **Y Angle**, or **Z Angle** input fields to rotate our components accordingly. *Figure 8.5* shows rotating one of the two components around its center of gravity by 45 degrees on the *X*, *Y*, and *Z* axes.

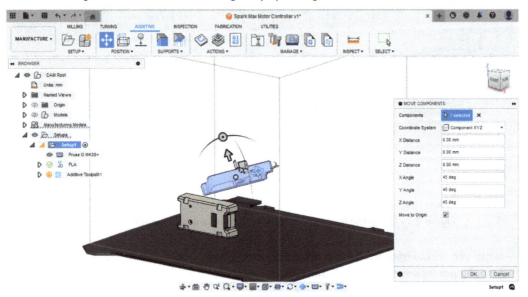

Figure 8.5 – Rotating a component using the MOVE COMPONENTS dialog

Once a component has been rotated, if we wanted to move that component a second time, after selecting it within the Move Components dialog, Fusion 360 automatically shows a local coordinate system at its origin, which is based on the last move operation we just finished.

If you look closely at *Figure 8.5*, you will notice that the coordinate system is set to **Component XYZ**. If we have already rotated a component once, the local coordinate system that will be displayed for that component will not match with the global XYZ axes. This will make it harder for us to translate a component in the global XYZ axes.

To address this issue, we can change our coordinate system from **Component XYZ** to **Machine XYZ**, which always matches up with the global coordinate system. After selecting the **Machine XYZ** coordinate system, we can translate our selected component(s) in the global X, Y, or Z directions with ease, as shown in *Figure 8.6*.

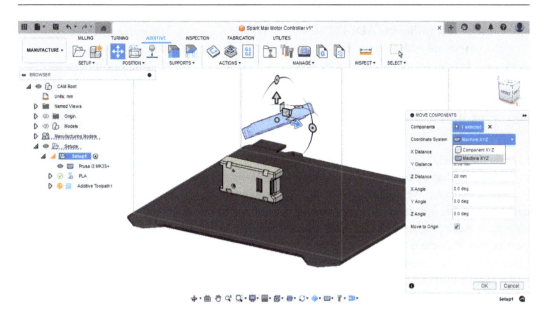

Figure 8.6 – Translating a component in the global Z direction using the Machine XYZ coordinate system

After orienting our parts with the Move Components dialog, we may want to quickly translate them closer to the bottom-left corner within our 3D printer's build volume. In such instances, after selecting the component we wish to translate, we can click the **Move to Origin** command, and our selected component will be automatically translated such that its bounding box's bottom - left corner coincides with the front bottom-left corner of our 3D printer's build volume, as shown in *Figure 8.7.*

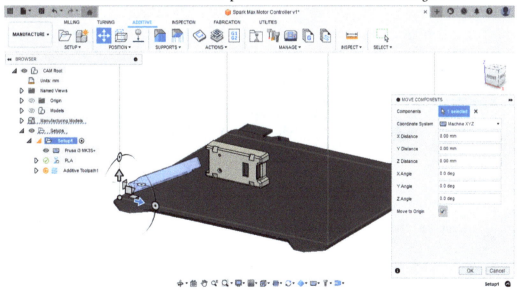

Figure 8.7 – Translating a selected component using the Move to Origin function

Another useful tool within the **ADDITIVE** tab's **POSITION** panel is **Minimize Build Height**. If we have a component with an irregular shape, or one that has been rotated such that it occupies a substantial Z height, we may want to search for an orientation where this component takes up the smallest possible Z height within our build volume. In certain additive manufacturing technologies such as **selective laser melting (SLM)**, **selective laser sintering (SLS)**, **multijet fusion (MJF)**, and binder jetting, it is beneficial to use the least amount of Z height so that we don't waste raw materials or print time.

Reducing the Z height of a given additive setup minimizes the number of layers that need to be 3D-printed. This has a positive impact on the overall printing time. Imagine if we were printing a build with 1,000 layers. We would have 999 intervals between the layers, where additive manufacturing technology-specific processes, such as recoating the build with raw material for the next layer, need to take place. Such processes can take ~10 seconds per layer interval depending on your 3D printer. If you could reduce the number of layers by rotating your components such that the build could be printed with half the number of layers (500 in this example), you could cut down the overall time you spend recoating the build by ~80 minutes. That is a substantial time saving that would have a positive impact on your manufacturing throughput and profitability.

The minimize build height command automatically rotates a selected component such that it occupies the least amount of space in the Z axis as shown in *Figure 8.8*. If we enter a non-0 input into the platform clearance field, this command will also leave a gap between the build plate and the part.

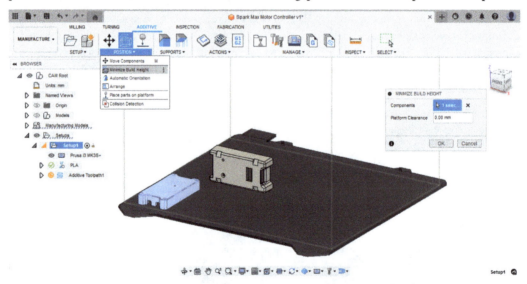

Figure 8.8 – Minimizing the build height of a component within an additive setup

The final manual positioning we will highlight in this section is the **Place parts on platform** command. As shown in *Figure 8.9*, this command offers us the ability to select either **Component**, or **Flat Face** within that component, using the **Type** dropdown. If we choose **Flat Face** from the dropdown, we can select a surface on a component and fill in **Platform Clearance** as desired, just as we described previously for the **Minimized Build Height** command. Upon clicking **OK** in the **PLACE PARTS ON PLATFORM** dialog, our selected component will rotate, and the surface we have selected will face the build platform. Please note that the **Flat Face** selection type only works with SOLID bodies and a planar face group on a MESH body cannot be selected as input for this command. *Figure 8.9* shows both the input and the output of this command on one of the two components we have in this additive setup.

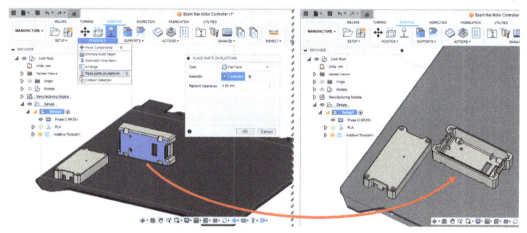

Figure 8.9 – Using the Place parts on platform command by selecting a flat face

In this section, we have covered three different manual part positioning commands that are available to us within an additive setup. Using these three relatively simple commands, we can translate and rotate our components so that they are within the build volume of our 3D printer. All the modifications to the position that we highlighted in this section were purely geometric. We did not consider any process-specific requirements as we were moving or rotating our components. In the next section, we will start to consider the requirements for various 3D-printing technologies and position our components accordingly.

Orienting Components Automatically

In the previous section, we talked about the various ways we can utilize Fusion 360 for positioning our parts within the build volume of our printer. However, all the translations and rotations we have shown up until now were explicit actions. Fusion 360 also offers tools to help us automatically orient our parts so that we can choose an orientation based on our 3D printer's specific technology in order to minimize material usage and increase our chances of a successful print.

To demonstrate how to automatically orient parts with Fusion 360, we will be using a model named `Connector Bracket,` as shown in *Figure 8.10*. This is a model we created in the previous chapter as a part of a larger assembly. In this chapter, we'll open this Fusion 360 design document and switch to the **MANUFACTURE** workspace. We will create an additive setup using an FFF printer (Prusa i3 MK3S +) and a PLA-specific print setting, as shown in *Figure 8.10*. We will also uncheck the **Arrangement** option within the **SETUP** dialog and click **OK** to generate the additive setup.

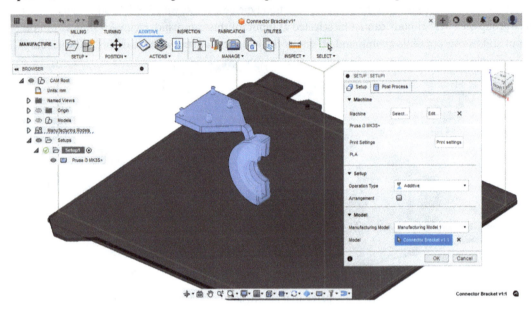

Figure 8.10 – Creating an additive setup with an FFF machine for a single component

Since we did not check the **Arrangement** checkbox while creating this additive setup, Fusion 360 does not automatically arrange our parts within the build volume of our printer. However, as the part was already positioned close to the origin, our part is automatically displayed within the build volume, and we can further position it as needed.

In this section, we will focus our efforts on the **AUTOMATIC ORIENTATION** command, which is located in the **ADDITIVE** tab's **POSITION** panel dropdown. Once we activate the **AUTOMATIC ORIENTATION** command, Fusion 360 will display a new dialogue with two tabs, as shown in *Figure 8.11*. The first tab, named **Parameters,** allows us to select a component on which to conduct the orientation study. In this example, we will conduct a single automatic orientation as this additive setup only contains one component. However, if we had multiple components in an additive setup, we could have conducted a unique orientation study for each component we wish to 3D print. After selecting the target component, we can edit the **Support Overhang Angle** input field and type in 45 deg as most FFF, SLA, DLP, and MPBF 3D printers are capable of creating overhangs up to 45 degrees without the need for support structures. The next checkbox, named **Support Bottom Surface,** allows us to add a platform clearance between our part and the build plate of the 3D printer when activated.

As this example is for an FFF printer, we don't need to activate it. However, if we were conducting this orientation study for an MPBF printer, we may want to activate that checkbox and enter a platform clearance of ~3 millimeters to account for the thickness of the cutting tool we will use to cut our parts off of the build plate once the printing process is complete.

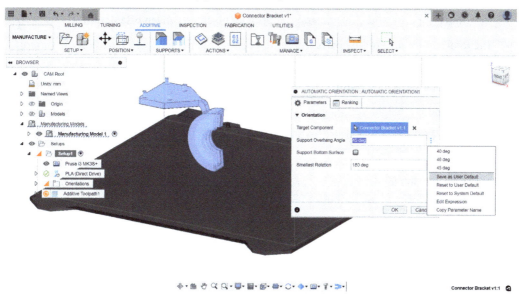

Figure 8.11 – Automatic orientation study for an additive setup with an FFF printer

Based on the inputs we have provided within the **Parameters** tab, Fusion 360 will generate various orientations of this component, and provide us with information about each orientation. To limit the long list of orientation results this command will generate, the **Parameters** dialogue allows us to adjust the minimum allowed rotation angle. By default, the **Smallest Rotation** input is set to 180 degrees. This means that the orientations we will get out of this study will be mirror images along a global plane such as XZ or YZ. If you reduce the smallest rotations input, Fusion 360 will produce more outcomes within the orientation results. However, this may take a slightly longer time to calculate.

My recommendation is to use the standard inputs within the **Parameters** tab and not deviate from the defaults if your additive setup is for an FFF printer. If you find yourself always needing to change the input for your orientation studies, you can left-click the icon next to the input represented by three vertical dots and select **Save as User Default** to set your custom input as the default, as shown in *Figure 8.11*. This way, the next time you utilize this command, the input of the dialog will be based on your desired preferences. This is a common way for you to customize the defaults used in most of the dialogs within the Fusion 360's **MANUFACTURE** workspace.

If you switch over to the **Ranking** tab within the **AUTOMATIC ORIENTATION** dialog, you will be able to control how Fusion 360 lists the outcomes of the orientation study, as shown in *Figure 8.12*. While generating a list of orientations for the selected component, Fusion 360 also calculates additional information such as the following:

- How much support would be necessary to print an object in a given orientation
- The total surface area on the parts where supports would be necessary based on the support overhang angle input
- What the bounding-box volume would be in a given orientation
- The maximum Z height occupied in a given orientation
- The height of the center of gravity in a given orientation

Using the dropdowns next to each of these ranking criteria, you can assign a priority to the values calculated by Fusion 360 for each item. As previously mentioned, the default ranking priority is the preferred ranking priority to use for both FFF and MPBF technologies.

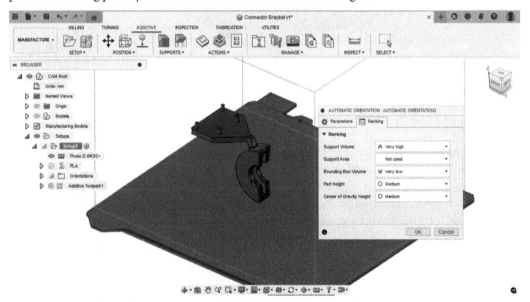

Figure 8.12 – Ranking criteria for an automatic orientation study

After conducting an orientation study and clicking the **OK** button in the **AUTOMATIC ORIENTATION** dialog, Fusion 360 adds a new line item called `Automatic Orientation1` within the **Orientations** folder in the browser for the active additive setup, as shown in *Figure 8.13*. Once Fusion 360 generates all the orientation outcomes, it will display the available orientation to choose from in a new tab called **Results** within the **ORIENTATION RESULTS** dialog. If you close this dialog by clicking **OK**, you can always come back to it by selecting `Automatic Orientation1` within the browser, right-clicking on it, and selecting **Orientation Results**. The results of an orientation study include a drop-down list ranking the orientation outcomes. By selecting the arrow key next to **Rank**, we can observe the results of the calculated support areas, support volume, bounding-box volume, overall height, and center-of-gravity height, as was previously outlined. The highest-ranking outcome, **Rank 1**, is the recommended orientation to print the target component on this FFF 3D printer.

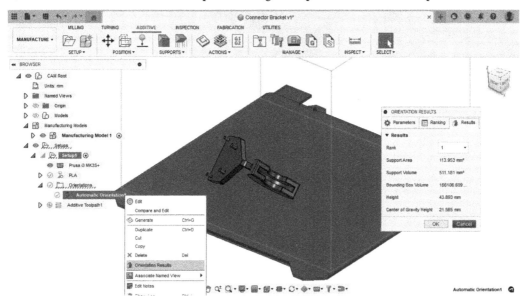

Figure 8.13 – Orientation results for an automatic orientation study

After selecting the desired orientation result using the **Rank** dropdown, we can click the **OK** button, and our component will rotate from its initial position to the orientation we have selected.

In this section, we mainly focused on conducting an automatic orientation operation and choosing the orientation result from a list of outcomes. However, it would also be beneficial to highlight the impact of this orientation on our selected component. To showcase the effect of this orientation, we can add a support structure to our component using the **SOLID VOLUME SUPPORT** command within the **ADDITIVE** tab, in the **SUPPORTS** panel. After selecting our component within the **GEOMETRY** tab of the dialog and clicking **OK**, Fusion 360 creates support structures for surfaces that have an overhang angle of less than 45 degrees. The outcome of the support generation action can be seen in *Figure 8.14,* along with the **SOLID VOLUME SUPPORT** dialog. The support structures Fusion 360 creates are displayed as blue objects on the canvas. We can see that this orientation produces a minimal support structure for our component so that it can be 3D printed using our FFF printer.

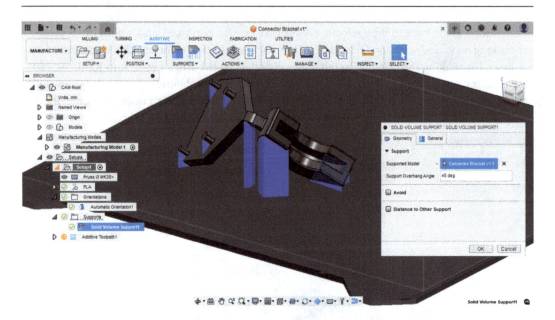

Figure 8.14 – Adding solid volume supports to a component after automatic orientation

It is important to remember that different 3D-printing technologies will require different parameters and ranking inputs when conducting automatic orientation studies. In the next example, we will highlight how we can use the automatic orientation functionality for choosing print orientations on an SLA/DLP 3D printer.

In this new example, we will use the same **Connector Bracket** document but create a new manufacturing model (`Manufacturing Model 2`), as shown in *Figure 8.15*. Then we will create a new additive setup (`Setup2`) that will utilize `Manufacturing Model 2` during model selection. This new setup will use an SLA printer named Prusa SL1S SPEED. We will also select any of the appropriate print settings for this printer from the available list of print settings.

After having created **Setup2**, we can focus on orienting the component. When determining the best component orientation with a resin 3D printer, we generally want to support the bottom surface and allow for a distance between the build plate and our part so we can generate a support that includes a base plate. We will cover the various types of support options based on 3D-printing technology in detail in *Chapter 10*. In *Figure 8.15,* this parameter is set to 7 millimeters. The main difference between the inputs for an automatic orientation for an FFF printer versus an SLA printer is in how we rank the outcomes. As shown in *Figure 8.15*, when ranking the outcomes, we do not rely on bounding-box volume, part height, or center-of-gravity height results. Instead, we look for the orientation that will result in the lowest support area required, while also trying to reduce the support volume as a secondary priority. The specific set of ranking priorities shown in *Figure 8.15* are my preferred ranking criteria when utilizing the automatic orientation tool for an SLA or a DLP 3D printer as these options tend to highly rank orientations that can be 3D printed with the smallest volume of support structures while producing quality models.

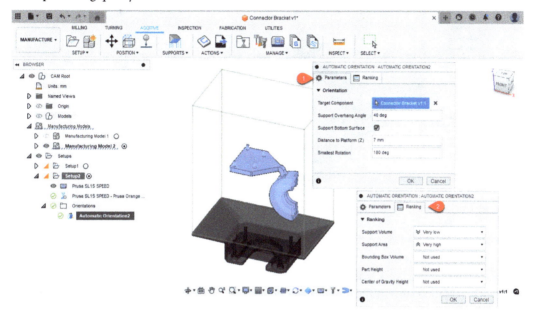

Figure 8.15 – Automatic orientation study for an SLA/DLP 3D printer

Based on the **Parameters** and **Ranking** criteria we set, Fusion 360 generates several options to orient this model. As you can see in *Figure 8.16*, the highest-ranked orientation (**Rank 1**) we get out of the orientation study has a very small area of support structures required (4 .07mm^3). This orientation will also require a relatively small amount of support volume (~94 mm^3). However, the outcome of the orientation study does not consider certain aspects of the 3D-printing process. To highlight the issue with this specific orientation, I made the part transparent and colored certain surfaces red, as shown in *Figure 8.16*. The red surfaces represent areas in the model where cupping can occur during the printing process. Cupping can cause print failures when 3D printing upside down using an SLA or DLP printer due to the pressure build-up in the trapped cavity.

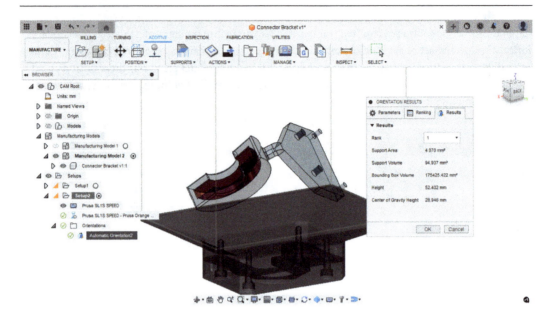

Figure 8.16 – Outcome of an orientation study for an SLA/DLP 3D printer

If we wanted to 3D print our model using this orientation, we would want to resolve the cupping issue by creating venting holes around the local minima of the cavity. Alternatively, we could choose a different orientation such as the one shown in *Figure 8.17* listed as **Rank 3**. If we print our part using this orientation, we will not have a convex portion of our component acting as a suction cup and trapping air during the 3D-printing process. This orientation will, however, result in a larger bounding-box height, which means a longer print time, and will require more support structures, meaning using more material during the print.

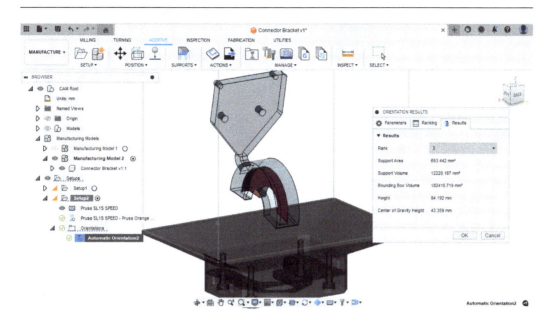

Figure 8.17 – Orientation to avoid the cupping effect for an SLA/DLP printer

After selecting the desired orientation using the **Rank** dropdown, we can click the **OK** button on **ORIENTATION RESULTS**, and our component will stay in the desired orientation. To demonstrate the impact of this orientation on our component, we can add support to this component by selecting the **BAR SUPPORT** command located in the **ADDITIVE** tab's **SUPPORTS** panel. Within the **BAR SUPPORTS** dialog, we can select our component, so that Fusion 360 can create support structures for all the surfaces that have an overhang angle less than the critical angle using **Medium** bars. Both the **BAR SUPPORT** dialog and the outcome of the support operation can be seen in *Figure 8.18* as a composite image. The support structures are displayed in blue in this image, and we can see that this orientation requires a number of bars to support the components during the 3D-printing process.

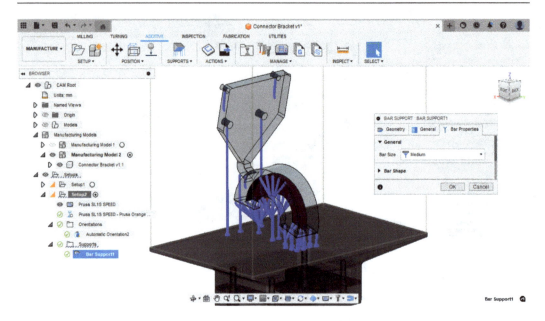

Figure 8.18 – Adding bar supports to a component after automatic orientation

When orienting our parts, there are numerous other factors to consider that depend on the given 3D-printing technology and the surface finish we require for our part. One key factor I would also like to highlight in this chapter is metal 3D printing using MPBF technology. As the MPBF process transfers energy from a laser to metal powder, parts that are 3D printed using this technology have a tendency to deform in the Z direction. If the deformation in the Z direction is above a certain value for a given layer, as fresh powder is being dispersed onto the next layer, parts could collide with the recorder blade, which would result in undesirable defects or print failures. Therefore, it is good practice to orient parts away from the direction of the recorder blade. *Figure 8.19* shows a design that leans into the path of the recorder blade. This may make the problem worse and cause the blade to collide with the part during the printing process due to the thermal expansion we just discussed. In such cases, we may want to orient our parts so that they lean away from the path of the recorder blade to avoid potential collisions with the blade.

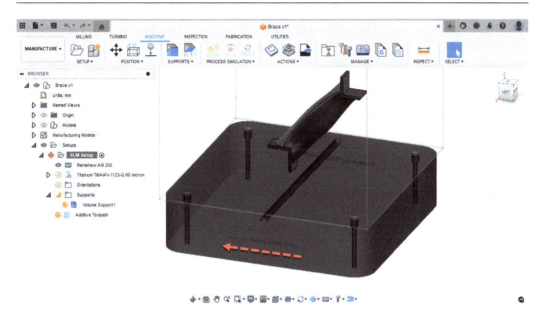

Figure 8.19 – Direction of the recorder blade versus component position

In this section, we talked about how to use Fusion 360 to automatically orient our parts. We also highlighted how to utilize the ranking mechanism within the **AUTOMATIC ORIENTATION** dialog to rank the outcomes of the orientation studies based on the needs of a given 3D-printing technology. We covered how to choose the top-ranking orientation as well as secondary orientations to improve the printing success without making further modifications to the model for SLA, DLP, and MPBF processes. In the next section, we will continue positioning our parts using additional automated tools such as additive arrange.

Arranging Components Automatically

Before utilizing 3D printing, companies often compare the cost per part against traditional manufacturing methods such as injection molding or CNC machining. When evaluating 3D printing versus injection molding for producing plastic parts, the economies of scale are on the side of 3D printing for prototyping and small production runs. It will generally be cheaper to 3D print a single component than injection-molding it. However, in small batch production, in order to decrease the per part cost of a given component, it is better practice to effectively 3D arrange parts within the build volume of capable hardware such as SLS and MJF 3D printers.

To arrange components within the build volume of a plastic powder bed printer, Fusion 360 offers two workflows. The first workflow can be accessed within the **DESIGN** workspace. The **ARRANGE** command located in the **SOLID** tab's **MODIFY** panel shown in *Figure 8.20* allows us to select one or more components and nest them within our build volume. In this section, we will demonstrate this arrangement functionality using an assembly of components named 3D arrange. This assembly consists of four components, each with a unique name and color (Red, Yellow, Blue, and Green). You may recognize these models from the previous sections. To generate this specific model, you can either use the **Derive** functionality and insert these components from their original source into a new Fusion 360 design document, or simply open the provided assembly file.

As mentioned previously, the first **Arrange** feature we will demonstrate is located in the **DESIGN** workspace in the **SOLID** tab's **MODIFY** panel. After activating the **ARRANGE** command, we can select the four components that will participate in this arrangement. They will show up as a vertical list of items within the **Objects** tab of the **ARRANGE** dialog, as shown in *Figure 8.20*.

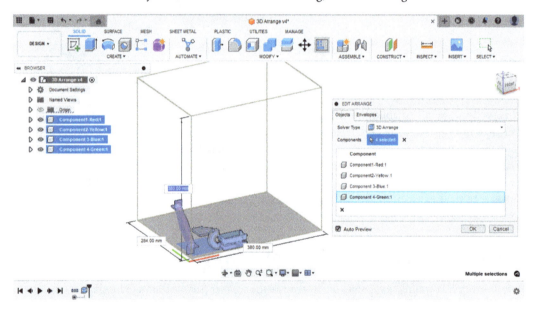

Figure 8.20 – Arranging components in 3D within the DESIGN workspace

If we switch to the **Envelopes** tab of the **ARRANGE** dialog, we can choose the global XY plane as the plane to arrange our components on. Later, we can enter the length, width, and height information, which represents the build volume of our 3D printer. In this example, we will be using 380 millimeters, 284 millimeters, and 380 millimeters respectively for the length, width, and height fields, representing the build volume of an HP Jet Fusion 5200 3D printer. This dialogue also allows us to perform partial arrangements. This means if we have more parts than we can fit into our build volume, Fusion 360 will arrange our parts so that those that can fit into the build volume are oriented accordingly, while parts that do not fit into the build volume stay in their initial position.

This dialogue also allows us to further control the spacing between parts, the clearance we may want to leave between our parts and the side walls of our build volume, and the clearance we may want to leave between our parts, the build plate, and the ceiling of the build volume.

If we activate the **Auto Preview** checkbox at the bottom of this dialog, Fusion 360 will automatically display the potential outcome of the arrangement while the dialog is still active, as shown in *Figure 8.21*. In this example, as we are in a parametric modeling paradigm, the **Arrange** function adds a feature in the timeline, as shown in the following screenshot. This means if we wanted to change any of the inputs of this arrangement, we could select the feature on the timeline, right-click it, and edit it with ease.

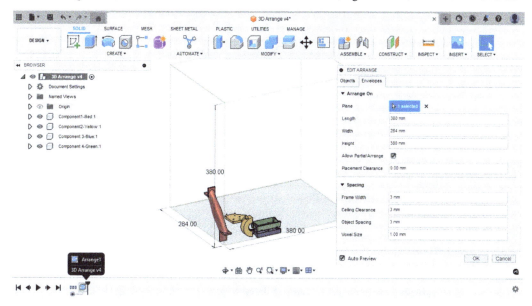

Figure 8.21 – Envelope-related inputs for 3D Arrange in the DESIGN workspace

The next method we can utilize for arranging our components within the build volume of our printer is located in the **MANUFACTURE** workspace. As shown in *Figure 8.22*, we can access the **Arrange** function within the **ADDITIVE** tab's **POSITION** panel, after having created an additive setup with an appropriate 3D printer and its relevant print setting.

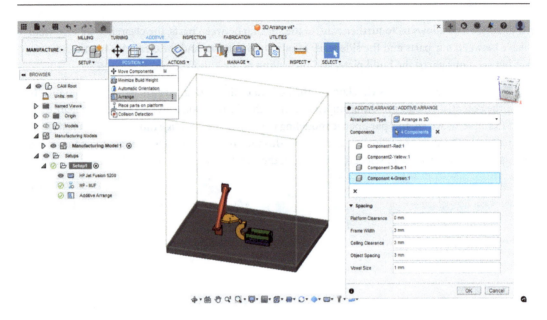

Figure 8.22 – Arranging components within the context of the 3D printer in the MANUFACTURE workspace

The **Arrange** command in the **MANUFACTURE** workspace is similar to the **ARRANGE** functionality we highlighted within the **DESIGN** workspace previously. Even though the **ADDITIVE ARRANGE** dialogue does not have the explicit option to allow for partial arrange, the outcome of this arrangement always arranges parts using this method.

The first difference between **Additive Arrange** in the **MANUFACTURE** workspace and the **Arrange** functionality in the **DESIGN** workspace is that **Additive Arrange** does not require you to explicitly enter the build volume dimensions. That input comes from the active setup you created and the printer you have selected. The second difference between these two arrangement methods is something you may have noticed while comparing the outcomes of the arrangements shown in *Figure 8.21* and *Figure 8.22*. If you look closely at *Figure 8.22*, you will notice that the arrangement outcome is such that the parts that are being arranged are concentrated around the center of the build plate. This is generally a more desirable outcome, as SLS, MJF, and binder jetting printers tend to have higher accuracy for part tolerance closer to the center of the build volume. This is because the center of the build volume can generally maintain its temperature more consistently. Therefore, it is desirable to place components close to the center of the build volume to avoid thermally induced deformations. It is generally better to avoid arranging parts near the walls of the build volume.

It is also important to have a basic understanding of the 3D arrangement solver Fusion 360 uses within the Arrange command. Regardless of whether you are using **Arrange in 2D** or **Arrange in 3D** for the **Arrangement Type** option in the **MANUFACTURE** workspace, as part of an additive setup, Fusion 360 utilizes an Outbox packer. This means that the arrangement of a given part starts with Fusion 360 creating a bounding box around the part. Later, Fusion 360 translates this box within the build volume without orienting the box or its contents in the process.

Outbox packers are known to pack parts in 2D and 3D really fast, but they don't produce the most effective packing-density results. If we want to achieve a better 3D packing density, we have a couple of options to try. The first option is to create copies of our parts in different orientations and see which orientation results in better arrangement density.

We will demonstrate this concept with the following example. However, to showcase arrangement density effectively, we will need more than four components so that we end up with a relatively full build volume. In *Figure 8.23*, you can see the same four components we worked on being duplicated using a circular pattern within the **DESIGN** workspace. In this example, we are creating 18 copies of the 4 components around the *Z* axis. This means each component will be rotated by 20 degrees around the *Z* axis and we will have a total of 72 components to arrange within our 3D printer's build volume.

> **Important note**
>
> SLS and MJF 3D printers generally produce parts with orthotropic material properties. This means that parts 3D printed using SLS and MJF are weaker in the *Z* direction of the print. Manufacturing engineers often identify the orientation in which to print their mission-critical parts and then arrange duplicate components in the same orientation without rotating them along their *X* or *Y* axes but allowing for *Z*-axis rotation if they are using an automated nester or manual arrangement tools.

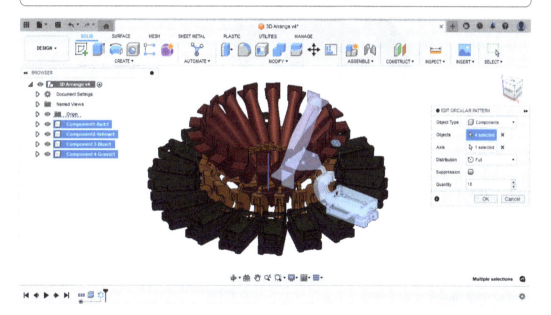

Figure 8.23 – Creating a circular pattern of components in the DESIGN workspace

Now we can switch to the **MANUFACTURE** workspace and create a new manufacturing model, as seen in *Figure 8.24*. After creating a new setup with the same HP Jet Fusion 5200 printer, we can arrange all 72 components in 3D using the Additive Arrange function. The following screenshot shows the outcome of arranging these 72 components within the build volume. As you can see, none of the components have been rotated beyond their original position. They were instead translated into place within the bounding box of the 3D printer. As this arrangement function utilizes a bounding box methodology, the components will not interfere with each other, and they will not experience interlocking.

Interlocking is a phenomenon where two separate components are arranged in such a way that after being printed, they cannot be separated. A common example of interlocking is the creation of a chain using two rings. If we wanted to print two rings, we would need to make sure that they are separate and do not interlock with each other. If they interlock, we cannot separate them after 3D printing them.

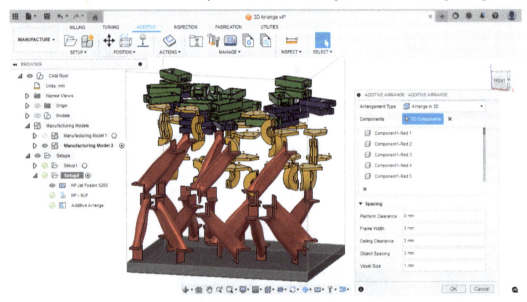

Figure 8.24 – Arranging 72 components in 3D within the MANUFACTURE workspace

As we were highlighting how to effectively arrange components within the build volume of our 3D printer above, we mentioned the concept of packing density. Packing density can be calculated while editing a manufacturing model used in an additive setup after all the part positioning actions are complete. After selecting the manufacturing model, we can right-click and edit. Right-click on the manufacturing model, select **Properties**, and access the **PROPERTIES** dialog for all the components in the build volume of our 3D printer, as shown in *Figure 8.25*. The **PROPERTIES** dialog shows us the total volume of components we have selected to inspect. Next, we will take a look at the height of the bounding box of our selected components.

In this example, the height is 337.289 millimeters. Using the information we have, we can perform a quick calculation and multiply the X and Y dimensions (380 mm and 284 mm respectively) of our 3D printer with the height value we obtained from the **PROPERTIES** dialog. Let's call this the printer utilization volume. Next, we can divide the volume of the parts by the printer utilization volume to calculate our arrangement density:

$$Arrangment\ density\ = \frac{Volume}{Bounding\ Box\ Height\ x\ (Printer\ Length\ x\ Printer\ Width)}$$

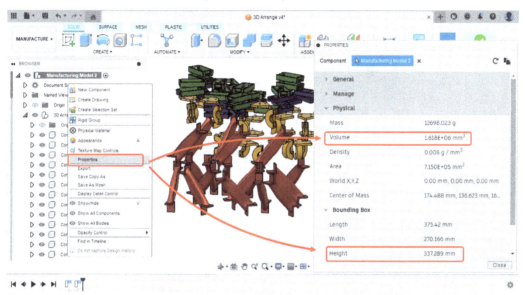

Figure 8.25 – Inspecting the manufacturing model for the height
of the bounding box of the arranged components

In this example, our arrangement density based on the printer utilization volume is approximately 4.4%. This is a good start, but we could have achieved a higher arrangement density if we used a rectangular pattern along the X or Y axes instead of a circular pattern around the Z axis while creating the duplicate components, as shown in *Figure 8.26*.

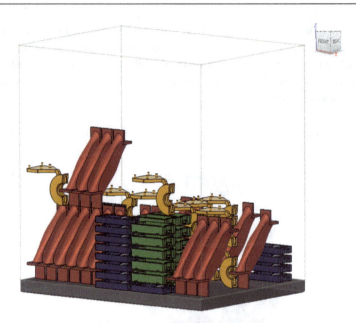

Figure 8.26 – Arrangement outcome of duplicate parts created using a rectangular pattern

Alternatively, we could have oriented all the components using the **Minimize Build Height** command in the **POSITION** panel of the **ADDITIVE** tab, as shown in *Figure 8.27*. Please note that the following screenshot demonstrates the outcome of this workflow after having created a new manufacturing model (Manufacturing Model 3) and a new setup (Setup3) that uses the new manufacturing model as input.

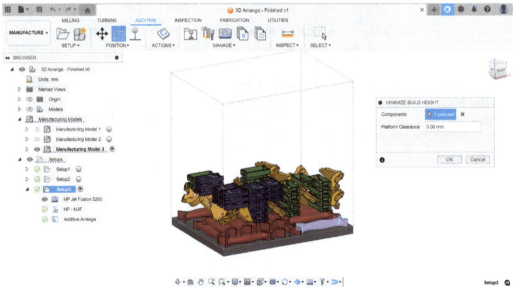

Figure 8.27 – Minimizing build height and 3D arranging all components

Once the parts are in an orientation where they occupy the least amount of *Z* height, we can perform the Additive Arrange functionality. *Figure 8.27* shows the outcome of Additive Arrange after having reoriented the parts using the minimized build height. You can see that this outcome has a much higher arrangement density based on the printer utilization. If needed, we could also manually move the components after the automatic **ADDITIVE ARRANGE** action is complete. Sometimes automated solutions cannot identify the best outcome, whereas, as manufacturing engineers, we can visually spot certain part orientations and positions that will benefit the overall arrangement density.

Summary

In this chapter, we talked about how to position and arrange our components for 3D printing. We started the chapter by showcasing how to translate and rotate our components manually using the move commands available within the **MANUFACTURE** workspace in the context of an additive setup. We also covered how to automatically orient our parts using explicit functions such as Minimize Build Height and Automatic Orientation. We talked about the various ranking options available within the **AUTOMATIC ORIENTATION** dialog so that the outcomes of an orientation study are ranked based on the needs of our additive technology. We ended the chapter by demonstrating how to arrange our components within the 3D build volume of our printer. In the final section, we covered how to use the automated arrange tools for existing orientations of components. We explained how we can orient and arrange our components within the **DESIGN** workspace, as well as the **MANUFACTURE** workspace in the context of an additive setup. We highlighted the benefits of using Additive Arrange in the **MANUFACTURE** workspace and talked about how to combine part orientation and part arrangement to achieve a higher arrangement density.

Now that we know how to arrange and orient our parts for 3D printing, we can start focusing on how to print each component with unique process parameters. In the next chapter, we will cover how to assign unique print settings to each component in order to modify how each component is printed, thereby impacting its physical properties.

9

Print Settings

Welcome to *Chapter 9*. In the previous chapters, we learned how to model for 3D printing and how to create an additive setup and position our components within that setup. We talked about effective ways to orient our parts and utilize 3D arrangement tools to get the most out of our 3D printer.

In this chapter, we will focus on print settings. We will highlight how to modify certain inputs within the print settings and their impact on the model that is being printed. We will also cover different print settings for different additive manufacturing technologies. Finally, we will touch upon how to assign a unique print setting on different parts within the build volume of our printer so that we can control how each part is printed.

Most people start their 3D printing journey by printing one component at a time. As they get more comfortable with 3D printing, they move on to printing multiple components in a single print. An important step to gaining that confidence relies on a better understanding of how to fine-tune our print profiles so that we can get reliable and successful prints from our 3D printer.

This phase of learning how to 3D print consistently will likely require some experimentation on your end with your specific 3D printer. The subjects we will cover in this chapter will make it easier for you to experiment with various print settings and dial in your print profiles for your specific printer and material.

In this chapter, we will cover the following topics:

- Overview of print settings for fused filament fabrication
- Overview of print settings for SLA and DLP printing
- Assigning unique print settings to bodies

In some cases, it may be necessary to edit the selected print setting to change a certain aspect of the printing process. For example, if you are experiencing print failures due to your parts not sticking to the build plate, you may want to add a skirt or a brim to improve print bed adhesion. If your 3D-printed parts are not solid enough, you may want to increase the infill density or the number of perimeters. In some cases, you may want to use a different print setting for a different part of the print. All of these and more will be explored in this chapter.

By the end of this chapter, you will have a better understanding of what makes a print setting in Fusion 360. You will learn that most print settings have multiple presets and you will understand the differences between those presets. You'll learn how to create your own presets and customize existing ones by changing certain print settings. You'll gain a solid understanding of key print settings and their impact on the additive toolpaths and slices Fusion 360 creates when generating the machine file. You will also learn how to assign print settings per body within a component and the impact on the additive toolpath.

Technical requirements

All topics covered in this chapter are accessible for Personal, Trial, Commercial, Startup, and Educational Fusion 360 license types.

Certain Fusion 360 additive machines and print settings, such as the **metal powder bed fusion** (**MPBF**) 3D printers and their respective print settings, are not available until gaining access to the **Manufacturing Extension**. As with all other Fusion 360 extensions, the Manufacturing Extension is not available for Personal use. As we will not cover topics specific to metal 3D printing in this chapter, you should be able to follow along with all the examples in this chapter with no issues.

If you have a commercial license of Fusion 360, to create an additive setup with an MPBF machine, you will also need access to the Manufacturing Extension (`https://www.autodesk.com/products/Fusion-360/additive-build-extension`).

Autodesk offers a 14-day trial for the Manufacturing Extension for all commercial users of Fusion 360. You can activate your trial within the **Extensions** dialog of Fusion 360.

The lesson files for this chapter can be found here: `https://github.com/PacktPublishing/3D-Printing-with-Fusion-360`.

Overview of print settings for Fused Filament fabrication

In this first section, we will focus our efforts on print settings that are specific to **fused filament fabrication** (**FFF**) additive manufacturing. To demonstrate how to access and edit print settings for FFF, we will be using a CAD model named **Knobs**, as shown in *Figure 9.1*. This part is actually the dial knob for a **Prusa i3 MKS+** 3D printer and is available to download as a mesh object on `printables.com` as one of the 38 components uploaded by Prusa Research as replacement parts for this 3D printer (`https://www.printables.com/model/57217-i3-mk3s-printable-parts/`). So, if you wanted to, you could download and print this part with a different colored filament or a different material and swap out your existing knob on your printer.

In this example, we will be using a solid (not the original mesh) version of this part. After opening the part in Fusion 360 and switching to the **MANUFACTURE** workspace, we can create an additive setup using the **Prusa i3 MKS+** printer and the **PLA (Direct Drive)** print setting, as shown in *Figure 9.1*. If we expand the print setting, we will notice that the preset that is assigned to this part is named **Normal**. If we want to edit a print setting or a preset, we can do so by right-clicking on the print setting within the active setup and selecting **Edit**. Another way is to go to **Print Setting Library** from the **MANAGE** panel. Within **Print Setting Library**, we can select the print setting we wish to customize from **Fusion 360 library**, and copy and paste it into either our **Cloud**, **Local**, or **Linked** library, as shown in *Figure 9.1*.

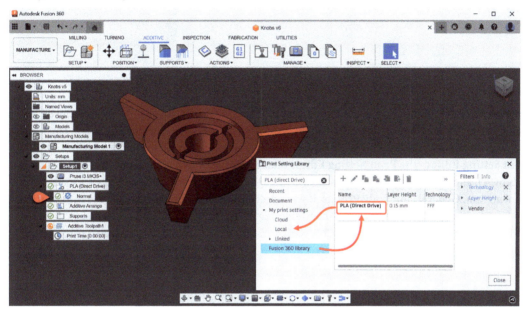

Figure 9.1 – Managing print settings within Print Setting Library

Once a print setting is saved under a folder within the **My print settings** section of **Print Setting Library**, we can select it, right-click on it, and edit it for further customization, as shown in *Figure 9.2*.

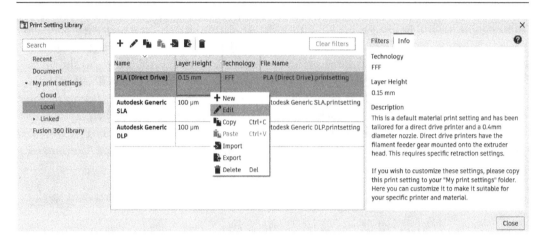

Figure 9.2 – Editing a print setting located in a Local folder

After selecting the **Edit** option, we will now see the **Print Setting Editor** dialog. We introduced this dialog in the previous chapter and talked about how to search for specific print settings and edit some of the key inputs. In this chapter, we will do a deeper dive and highlight all the options available within **Print Setting Editor**. The **Print Setting Editor** dialog consists of two main areas. The left portion is its browser, and the right portion allows us to change the input. The FFF print settings are made up of **Information**, **General**, and **Body Presets**. If you select **Information** on the browser, the **Information** input field will be displayed on the right-hand side. *Figure 9.3* shows our first modification to this print setting, which is the addition of the - CUSTOMIZED suffix to the name of the selected print setting.

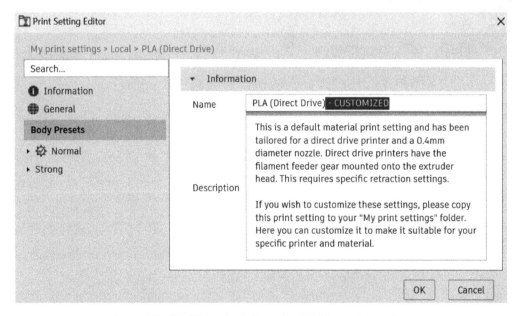

Figure 9.3 – Modifying the Information field for a print setting

If you select **General** on the browser, you will be able to modify the input for **Global Settings**, **Nozzle Priming Settings**, and **Additional G-Code Settings**.

> **Important note**
> Please keep in mind that not all 3D printers are capable of managing all these inputs. For example, if you have a printer without a heated print bed, the **Print Bed Temperature** field will have no effect on your print.

General

The **General** section of an FFF print setting is where you can modify some of the most common input fields when customizing your print profiles. *Figure 9.4* shows all the input fields you can customize within the **General** section for an FFF print setting. This section is divided into three categories: **Global Settings**, **Nozzle Priming Settings**, and **Additional G-Code Settings**. In the following subsection, we will go over some of the critical input fields within each of these categories and their use cases.

Global Settings

The **Global Settings** subsection located within the **General** section of **Print Setting Editor** is probably the most important part of the print settings. This is where you can edit things such as the build plate temperature and control the height of each layer as well as your first layer. You can also adjust how fast your extruder will move during printing and decide whether you want to print your parts as they're designed or only their shell using the vase mode. Let's take a moment and go through each one of these options and their role in generating the additive toolpath:

- **Print Bed Temperature (Temp C)**: This is a material-specific input that sets the temperature of the print bed before starting the printing process. A heated bed helps with print bed adhesion of the first layer. In *Figure 9.4*, the print bed temperature is set to 60°C, which is the recommended bed temperature for printing with a PLA filament. If you were using ABS, this input would be 85°C.

- **Layer Height (mm)**: This input specifies the height of each layer. You can reduce the value for better print quality or increase it for a faster print. The upper and lower limits of layer height are dependent on your nozzle size.

- **Layer 1 Height (mm)**: This input specifies the height of the first layer of your 3D print. Typically, this input should be 50 to 75% of your nozzle diameter to achieve a successful print bed adhesion on the first layer.

- **Travel Speed (mm/s)**: This input specifies the speed with which your extruder will be moving when traveling between points while not extruding filament. In general, you would want this input to be as close to the maximum speed your printer can handle to reduce print time.

- **Spiral Vase Mode:** When activated, this will print the part in your active setup using one continuous spiral motion to create a hollow, single-walled object. This mode results in fast print times and avoids any travel and retraction moves during 3D printing. If enabled, many of the inputs within the **Body Presets** section will be ignored as they will no longer be applicable.

Nozzle Priming Settings

The **Nozzle Priming Settings** subsection located within the **General** section of **Print Setting Editor**, as shown in *Figure 9.4*, controls all the options related to printing a prime line before the extruder starts working on your part. Let's highlight some of the key features you can control around the prime line:

- **Nozzle Priming:** Enabled by default, this option will print a single line at the front, the left, or the right-hand side, or the back of your print bed. The location of the priming line will depend on your part arrangement in the additive setup and the availability of unused space on the print bed. Printing a prime line prepares the nozzle for 3D printing. By printing a prime line, you ensure that the filament is extruding correctly before printing your part.

- **Prime Edge Clearance Distance (mm):** This input controls how far away the prime line should be from the border of your build plate.

- **Prime Maximum Length (mm):** This input controls the maximum length of your prime line. I recommend increasing this number to match the *X* or *Y* dimension of your 3D printer's build plate.

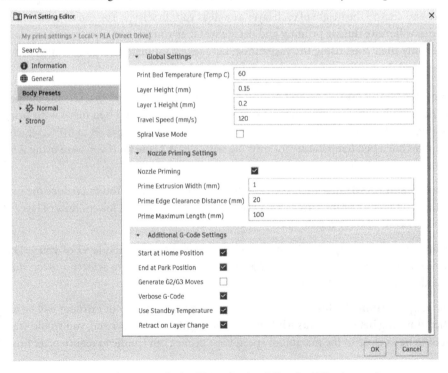

Figure 9.4 – The General tab of Print Setting Editor for FFF print settings

Within the **General** section of **Print Setting Editor**, you can also control additional G-code settings, as shown in *Figure 9.4*. By default, any G code you generate always starts at the home position and ends at the park position. If your 3D printer has support for it, you can also generate G2 and G3 moves by checking the corresponding checkbox in this dialog. This action would help create a G-code file with single G2/G3 rows of text instead of multiple G1 rows of text to represent an arc-shaped toolpath in a given slice. If you are not familiar with it, G2 is for a clockwise arc, and G3 is for a counterclockwise arc. Using a single row of G code to represent the arc is more efficient for file size and results in higher-quality/smoother 3D prints.

Body Presets

The next section within **Print Setting Editor** for an FFF print setting is **Body Presets**. By default, all print settings included in Fusion 360's **Print Setting Library** have two body presets. They are named **Normal** and **Strong**. If you select one of the presets, you can change its name and description on the right-hand side, as shown in *Figure 9.5*. In this figure, we are changing the preset name from **Normal** to Medium.

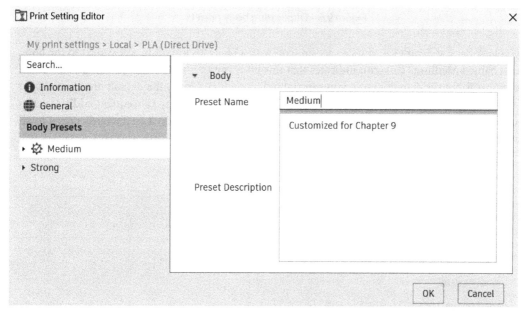

Figure 9.5 – Changing a body preset name and description

We can also right-click on a body preset and duplicate it. After duplicating a body preset, you will see a new entry at the bottom of the browser of **Print Setting Editor**. In *Figure 9.6*, you will see that a new entry, **Medium 1**, was added after duplicating the body preset named **Medium**.

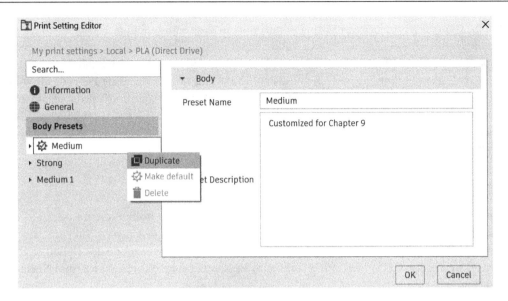

Figure 9.6 – Duplicating body presets

If you look back at *Figure 9.6*, you can see an icon with a blue checkmark in a gear next to the body preset named **Medium**. This icon indicates that this body preset is the default preset. If you select a preset that is not the default preset, you can right-click on it and make it a default preset, as shown in *Figure 9.7*. If you were to make this change to this print setting, any subsequent additive setup you create using this print setting would automatically assign **Medium 1** as the body preset to all components within that setup.

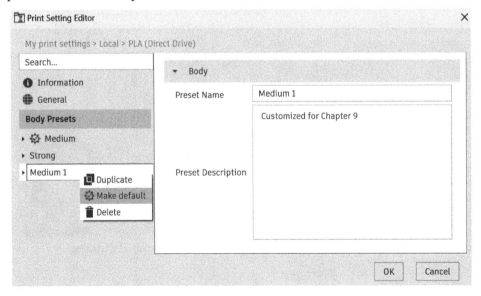

Figure 9.7 – Setting a body preset as default

If we expand a given body preset by selecting the arrowhead to the left of it, as shown in *Figure 9.8* for the body preset named **Medium**, we can see a number of entries that control different aspects of the **Body Preset** setting, such as **Extrusion**, **Shell**, **Infill**, and several others. Next, let's dive into some of these settings and explain what they are.

Extrusion

The input fields within the **Extrusion** subsection of a given body preset control which extruder is used for printing bodies within an additive setup and certain settings associated with that extruder. If we are creating an additive setup using a 3D printer with a single extruder, we will only see the subheadings called **Extrusion** and **Extruder 1** when editing the print setting for that setup, as shown in *Figure 9.8*. Only when editing a print setting associated with a multi-extruder printer can we see additional extruder subheadings in this dialog.

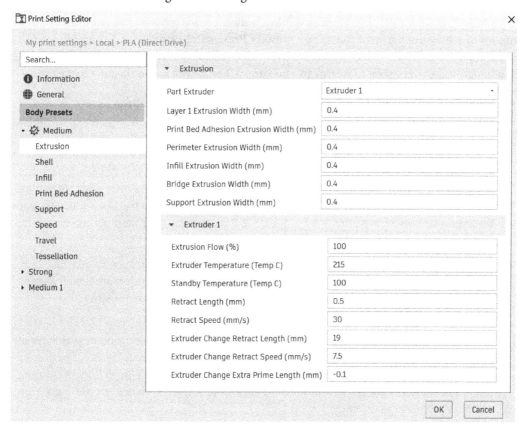

Figure 9.8 – Extrusion settings for the default body preset of a PLA (Direct Drive) print setting

The **Extrusion** subheading controls the width of the first layer as well as print bed adhesion with dimensions for brim, skirt, and raft. It also has inputs to modify the extrusion width of key toolpath types such as **Perimeter**, **Infill**, **Bridge**, and **Support** structures. Within the **Extruder 1** subheading, you can modify another critical input, **Extruder Temperature (Temp C)**. In *Figure 9.8*, the value shown for this input is 215°C, which is the appropriate temperature to 3D print a PLA filament. If we were to edit a print setting for a different material such as ABS, this value would be prepopulated with the appropriate temperature of **250°C**.

In the next section, we will learn how to control various settings related to 3D printing the outer surface of your parts. We will highlight how to modify the number of perimeters and top and bottom layers, as well as how to avoid surface imperfections at the beginning of each layer.

Shell

When slicing a model for FFF, one of the first things a slicer does is create a shell of the model. The options we have within the **Shell** subsection of **Body Presets** in **Print Setting Editor** allow us to have full control over these options to create and slice that shell. Let's take a look at some of the most common settings you may want to modify to generate a 3D print with a smooth outer surface:

- **Number of Perimeters**, **Number of Top Layers**, and **Number of Bottom Layers**: As shown in *Figure 9.9*, within the **Shell** section of a given body preset, you can control how many lines of toolpath you want to have around the perimeter of your component, as well as how many top and bottom layers you want the slicer to include. The higher the number of perimeters, top layers, and bottom layers you include when slicing your model, the stronger it becomes along those sections. The top and bottom layers are unique in that the slicer adds 100% infill to those layers, making them fully solid. *Figure 9.9* shows a model that has two perimeters. For the layer displayed in the figure, layer 121, the red line represents the outer perimeter and the green line represents the inner perimeter.

- **Randomize Z-Seam Alignment**: When activated, this setting allows you to distribute any imperfections you may have on the perimeter of your model so that they are not all aligned with the *Z* axis. If unchecked, your printed model may have a seam/zipper effect.

- **Print Inner Perimeters First**: When 3D printing a given layer, the extrusion has to start somewhere. Slicers tend to start extruding each layer at a point along the outer perimeter of your model. When active, this setting forces the extrusion to start along the inner perimeter of a given slice. In *Figure 9.9*, the inner perimeter is represented with green lines. Activating this checkbox forces any imperfection at the beginning of a given layer to be hidden within the inner perimeter of the body.

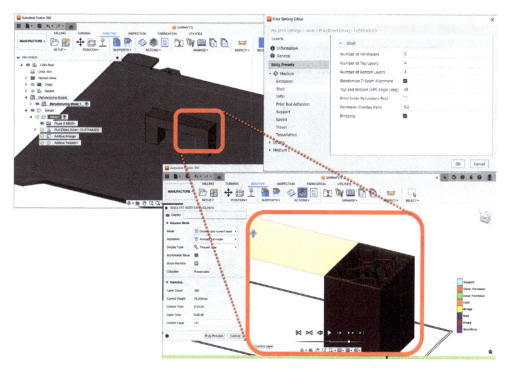

Figure 9.9 – Shell settings for the default body preset of a PLA (Direct Drive) print setting

By modifying these options, you can slice and print the shell of your models with your FFF printer so that you end up with your desired surface quality and strength-to-weight ratio for your parts. In the next subsection, we will learn how to modify the options for the interior of our parts.

Infill

When slicing a model for FFF, the term *infill* represents the internal section of the 3D-printed part. This internal volume can be manufactured using many different shapes. The purpose of the infill is to minimize the part's weight while keeping it as strong as possible and reducing printing time. There are many different types of infill patterns Fusion 360 offers within the **Infill Pattern** dropdown located in the **Infill** subheading of a given body preset, as shown in *Figure 9.10*. The same figure also shows three components being printed with a 25% density **Gyroid** infill. Notice how the internal sections of all the parts are not printed solid with filament, but rather, are being manufactured mostly hollow with gyroid-style toolpaths.

> **Important note**
> A recent study on all infill patterns available within Fusion 360 (as documented in the blog at `https://www.autodesk.com/products/fusion-360/blog/how-to-build-sustainable-3d-printing-infill-practices/`) concluded that the **Gyroid** infill is the most sustainable infill type as it prints the fastest while minimizing filament consumption.

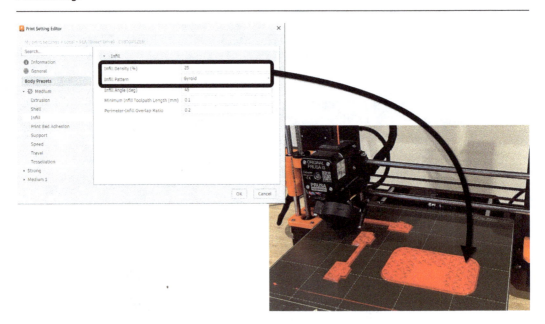

Figure 9.10 – Infill settings for the default body preset of a PLA (Direct Drive) print setting

By modifying these options, you can slice and print the insides of your models with your FFF printer so that you end up with a lightweight part that you can print faster while consuming less material and energy. In the next subsection, we will learn how to modify the options to achieve good adhesion to the build plate while printing our first layer.

Print Bed Adhesion

One of the most important considerations when slicing models for 3D printing with FFF is the topic of print bed adhesion. A key failure when 3D printing occurs as a result of issues around parts not sticking to the print bed. Fusion 360 offers several tools to overcome this potential failure within the **Print Bed Adhesion** subheading of **Print Setting Editor** for a given body preset.

Within the **Print Bed Adhesion** subheading, you can control how many layers to disable the cooling fan for so that the first few layers can be printed without the cooling effect the fan provides. In addition, you can select from one of three **Print Bed Adhesion** options that Fusion 360 provides:

- **Skirt/Brim**: Even though **Skirt** and **Brim** have explicit options to choose from within the **Print Bed Adhesion type** dropdown, they are effectively going to give you the same toolpath when slicing your model. Both a skirt and a brim will follow the perimeter of the first layer of your part and will lay down additional lines of toolpath around the outer perimeter. You can control how many loops of skirt or brim you want to add around the perimeter of the body. If you choose **Skirt** for **Print Bed Adhesion type**, you have one additional control, which is the distance between the skirt and the perimeter of your body. If you enter 0 in this field, you will essentially generate a brim. As these two are almost interchangeable options, when simulating the toolpath, Fusion 360 displays **Skirt** and **Brim** as the same color in the legend, as shown in *Figure 9.9*.

- **Raft**: When 3D printing with materials that are more prone to warping and causing bed adhesion issues, you may choose to use a raft, as rafts help stabilize models and create a strong base on which you can print the rest of your part. If you choose **Raft** for **Print Bed Adhesion type**, Fusion 360 will activate additional **Raft** settings, as shown in *Figure 9.11*. The default settings for a raft will result in a three-layer **Raft** object with varying infill densities for the bottom layer, middle layer, and top layer. You can also control the border offset of your raft from the perimeter of the first layer of your model.

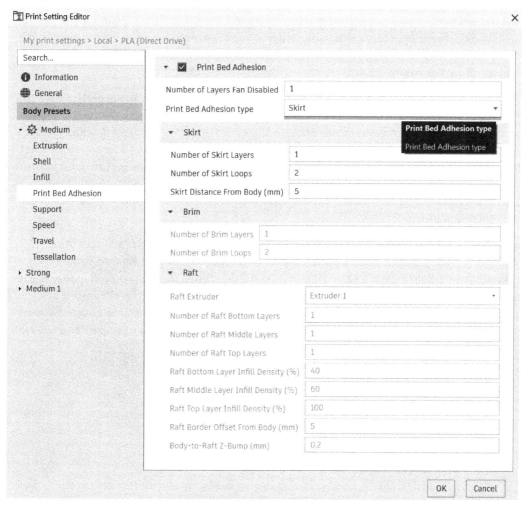

Figure 9.11 – Print Bed Adhesion settings

By modifying these options, you can achieve a good adhesion to the build plate while printing your first layer. In the next subsection, we will learn how to modify the options to slice and generate toolpaths for support structures you can generate within Fusion 360 for FFF additive setups.

Support

Even though it is possible to manufacture certain overhanging faces with bridging techniques, in most cases, we rely on generating support structures when an overhang angle is more than 45 degrees so that our part's down-facing surfaces can be printed successfully.

When preparing our models for printing with FFF, Fusion 360 offers three support types: **volume supports, bar supports, and setter** supports. Setter supports are mainly used for 3D printing with a metal filament or 3D printing using metal binder jetting technologies to help support the part during the sintering process. In order to generate a setter support, we need access to the Manufacturing Extension. We will not cover setter supports further here as they are not applicable to any of the examples we are showcasing.

Within an additive setup with an FFF machine, you can generate volume and bar supports and we can mix and match support types, meaning we can support one face with volume supports and another face with bar supports. Therefore, Fusion 360 also allows you to control how you slice and generate toolpaths for these support structures. You can control which extruder is used for a given support structure as well as how many perimeters are utilized in the toolpath generation for bar versus volume supports. You can also control the number of top and bottom layers and how much infill density is required when generating the toolpath for a support structure within the **Support** subheading of the body preset, as shown in *Figure 9.12*.

One setting I would like to highlight in this section is **Support-to-Body Horizontal Gap (mm)**. This input controls how much horizontal space Fusion 360 leaves between your support structures and any vertical wall on the perimeter of your part. If you increase this number, you will have a bigger clearance between your support structures and your parts along the vertical walls, eliminating possible bonding between the part and the support structure. This setting does not control the distance between the support and the down-facing surfaces of your part. We will cover how to control such options in *Chapter 10*, in which we talk about how to generate support structures.

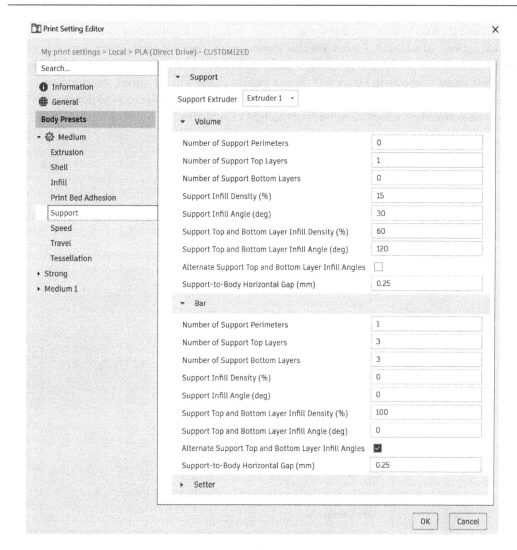

Figure 9.12 – Support settings

By modifying these options, you can control exactly how Fusion 360 slices the volume and bar supports that are commonly used in FFF setups. In the next subsection, we will learn how to modify the options to control the speed with which to extrude material within Fusion 360 for FFF additive setups.

Speed

One of the main drawbacks of FFF is the time it takes to print large parts. This is why you may want to modify the **Speed** settings within the **Body Presets** section of **Print Setting Editor** for your selected print setting. These settings are very self-explanatory. You can control how fast to execute various extrusion types such as first layer, perimeter, infill, support, bridge, and raft.

If you activate the **Acceleration** checkbox, you can also control the **Acceleration** inputs for travel and extrusion actions. In addition, Fusion 360 also allows you to control the maximum instantaneous velocity change while printing. This setting is called **Jerk** and is located at the bottom of the **Speed** subheading menu, as shown in *Figure 9.13*. Before making changes to the **Speed**, **Acceleration**, and **Jerk** settings, you should be aware of the specifications of your 3D printer so as to not exceed the velocity and acceleration values based on your machine's limitations.

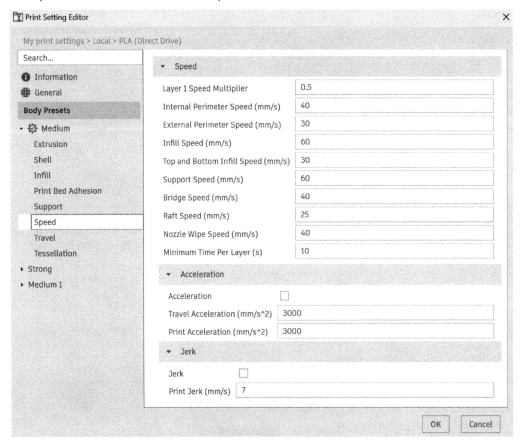

Figure 9.13 – Speed settings

By editing the options within the **Speed** subsection of **Print Setting Editor**, you can have full control over the velocity, acceleration, and jerk behavior of your 3D prints. In the next subsection, we will cover how to modify the speed with which the nozzle travels when not extruding material.

Travel

Travel lines are the movement of the nozzle without extrusion during FFF 3D printing. The **Travel** subheading within **Body Presets** controls the various inputs for travel during slicing.

If you are experiencing issues with your extruder hitting your print or generating blobs on the surface of your model due to oozing material, you may want to activate **Rapid Z Lift**. This setting will elevate your extruder by a set distance when traveling from one section of your model to another while printing. This should prevent your nozzle from hitting the previously extruded sections of your part, help reduce blobs, and decrease print failures. However, this option will add extra time to your prints. Therefore, you may also want to enable **Rapid Z Lift Diagonal Move** to minimize the time you will spend when performing **Rapid Z Lift** actions. Fusion 360 also performs a **Nozzle Wipe** action by default for all print settings. This command will drag the nozzle a set distance over the extrusion that was just performed, with the extruder off, to help wipe the nozzle and potentially minimize stringing.

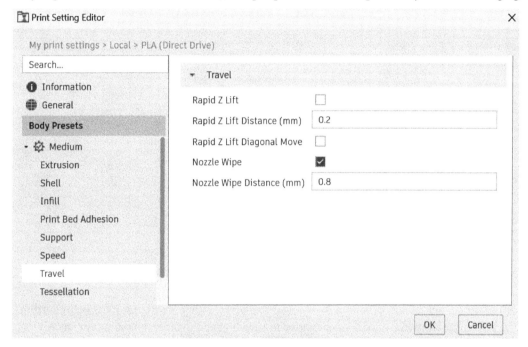

Figure 9.14 – Travel settings

By modifying these options, you can achieve an overall reduced print time and increase your chances of a successful print by avoiding crashes between your extruder and your parts. In the next subsection, we will learn how to modify the options to tesselate parts within Fusion 360 before slicing them within an additive setup.

Tessellation

When we are working with solid bodies in an additive setup, in order to slice them and create toolpaths, Fusion 360 first tessellates those bodies and turns them into a mesh representation. In order to go through this conversion, it relies on the values shown within the **Tessellation** subheading of a given body preset. We already covered how to tessellate models in great detail in *Chapter 5*. The **Tessellation** settings shown in *Figure 9.15* correspond to the same **Tessellation** settings we covered within the design workspace of Fusion 360 when converting a solid body to a mesh body. By default, the **Tessellation** settings used for all print settings are such that Fusion 360 will create a high-quality mesh so that the mesh body used for slicing our models is an accurate representation of our solid body.

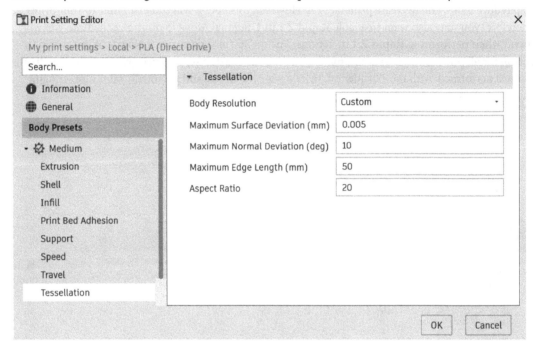

Figure 9.15 – Tessellation settings

Now that we've highlighted how to make changes to a print setting within our local **Print Setting Library**, it is time for us to assign this print setting to our active additive setup. As shown in *Figure 9.16*, we can select the **ADDITIVE** setup, right-click, and edit it. When we are editing the setup, we can choose a new print setting. We will have to navigate to the **Local** folder, within the **My print settings** section of **Print Setting Library**, and choose the print setting we recently customized called **PLA (Direct Drive) - CUSTOMIZED** and press **Select** and **OK** to finish editing the setup.

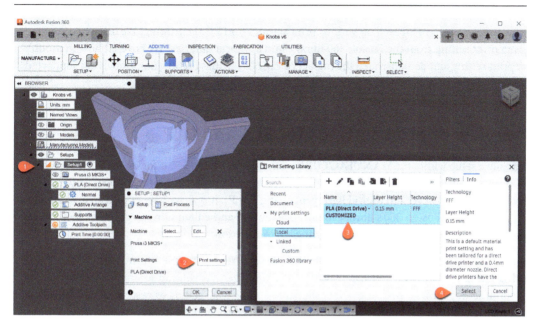

Figure 9.16 – Selecting a custom print setting within an additive setup

Once we have selected the print setting that we modified in the previous sections of this chapter, you will notice that the active setup now displays a new entry named **PLA (Direct Drive) - CUSTOMIZED**, and under that entry, a body preset named **Medium** is shown, indicating that our part will be sliced using that body preset.

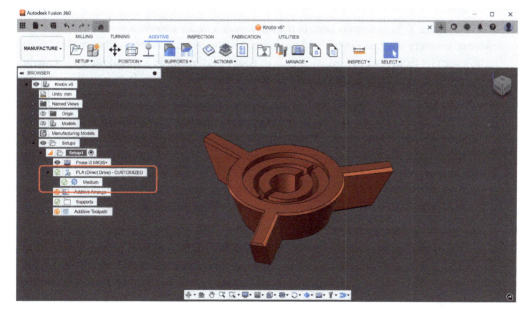

Figure 9.17 – Observing the selected print setting and body preset in the browser

In this section, we have focused on various print settings for FFF 3D printing. We discussed copying a given print setting from the Fusion 360 library into our local libraries. We covered how to modify that print setting and demonstrated the impact of various settings on how our models are supported and sliced, and how the toolpath is generated. In the next section, we will start looking at print settings for other 3D printing technologies.

Overview of print settings for SLA and DLP

An additive setup within the **MANUFACTURE** workspace of Fusion 360 is for more than just desktop FFF printers. By using Fusion 360, you can set up your prints for resin-based printers, which utilize **stereolithography (SLA)** and **digital light processing (DLP)**. Just as we have copied an existing FFF print setting from Fusion 360's library to our local library and edited it in the previous section, we will do the same for SLA and DLP print settings in this section.

Figure 9.18 shows a print setting named **Autodesk Generic DLP** copied from Fusion 360's **Print Setting Library** to **My print settings | Local** to be customized. As you can see in the same figure, the DLP print setting browser structure is very similar to an FFF print setting browser structure. This DLP print setting has three sections, named **Information**, **General**, and **Body Presets**. *Figure 9.18* shows the **General** subsection of the print setting where you can configure various build parameters. The main editable field within **Configure Build Parameter (EBPA) Settings** is **Layer Thickness**. Even though the units are not shown for layer thickness in this dialog, they are in microns and can be seen when selecting the print setting in the **Information** panel in **Print Setting Library**.

Within the **PNG Export** subsection, the first six inputs are critical in order to generate the desired PNG files with the correct size:

- **Min X** and **Min Y**: These inputs control the minimum X and Y coordinates of the image to be captured on every layer within the build volume. A value of 0 means you will capture all your images starting from the bottom-left corner of your build platform. I would not recommend changing these inputs if you want to capture the entire build.

- **Size X** and **Size Y**: These inputs control the maximum X and Y coordinates of the image to be captured on every layer within the build volume. The units for these two inputs are in millimeters. You definitely want to change these inputs to match your printer's printable X and Y dimensions. For example, if your printer has a printable build volume of 320 millimeters by 240 millimeters, you will want to type those numbers in for the **Size X** and **Size Y** input fields, respectively, within this print setting before slicing your models. Otherwise, the images you will export from Fusion 360 will be only a subsection of the entire build area.

- **Image Size X** and **Image Size Y**: These inputs control the pixel size of the images you will export. Typically, these numbers should be a multiplier of the numbers you have entered in **Size X** and **Size Y**. For example, if you have a printer with a printable build area of 320 by 240 millimeters, you can enter an **Image Size X** and **Image Size Y** input of 640 and 480 pixels, respectively. If you enter values that are not multipliers of the **Size X** and **Size Y** values, you will get distorted images during slicing and, therefore, distorted objects during the printing process.

- **Flip Image**: Depending on the placement and the model of the projector (LCD versus DLP) your 3D printer is using, you may need to export mirror images for all your slices so that you don't 3D print a mirror image of the object.

- **PNG Filename Prefix**: Different 3D printers have different requirements for how PNG files should be named. Using this input field, you can customize how each PNG is named by adding a prefix. If you simply keep the existing prefix of image_, the resulting PNG files will be named image_000.png, image_001.png, and so on.

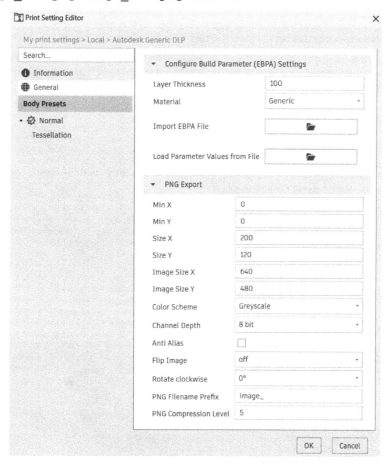

Figure 9.18 – Autodesk Generic DLP print settings

After creating an additive setup with an Autodesk Generic SLA/DLP printer and the **Autodesk Generic DLP** print setting, if you select **Create Machine Build File** located within the **ACTIONS** panel of the **ADDITIVE** tab, you will be able to export each slice as a PNG file with your specifications, compressed as a single ZIP file. You can then further edit the ZIP file to include any additional TXT or XML files your printer may need before transferring it to your 3D printer for manufacturing your models.

The next print setting I would like to highlight is the **Autodesk Generic SLA** print setting. If you select this print setting from Fusion 360's **Print Setting Library** and copy it over to your local library, you can edit the print setting in **Print Setting Editor**, as shown in *Figure 9.19*. Much like the **Autodesk Generic DLP** print setting and all the FFF print settings, the **Autodesk Generic SLA** print setting contains subsections named **Information**, **General**, and **Body Presets**. The **General** subsection is where we can edit **Layer Thickness** in microns. This print setting allows us to create a machine build file that generates either a **stereolithography contour** (**SLC**) file or a ZIP file. This ZIP file contains one or multiple SLC files. SLC files are files that contain slice information for an SLA printer. Within the print settings, you can control whether you want to create unique files for each part and/or unique files for each of the support structures you may have generated during the additive setup. You can also choose to use the name of the component as the filename when exporting SLC files. Fusion 360 automatically includes a postfix for the name of the support structure. The default postfix is _s. If you use the default settings and export an SLC ZIP file, it will contain the following two SLC files:

- `Component1_Body1.slc`

- `Component1_Body1_s.slc`

This is assuming that your additive setup contains a single body component named `Component 1` and that you have added a support structure to it.

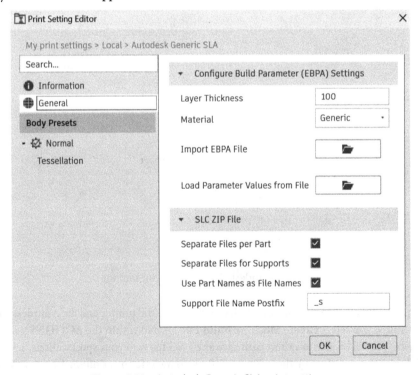

Figure 9.19 – Autodesk Generic SLA print settings

In addition to the generic SLA and DLP print settings, Fusion 360 has three machine vendor-specific print settings for resin printers. The first one is for **Anycubic Photon Mono**. If you access **Print Setting Library**, you can search for this print setting and visualize its inputs within **Print Setting Editor** after copying it to your local library. As shown in *Figure 9.20*, this print setting looks very similar to a generic SLA or DLP print setting but with one main difference. The **General** subsection of this print setting includes inputs for four parameters related to how a given material should be sliced. You can control inputs such as exposure time and the number of bottom layers for a given additive setup. You can also control how much time the printer should pause for curing each layer in seconds.

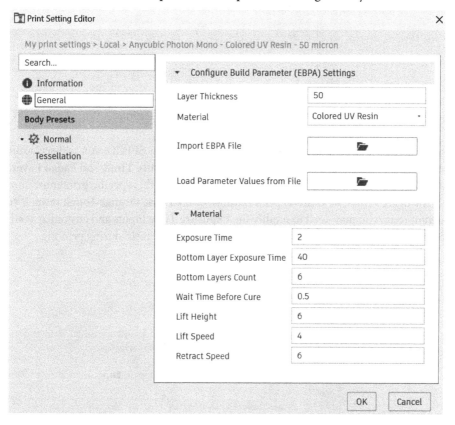

Figure 9.20 – Anycubic Photon Mono print settings

If you have a DLP 3D printer from the Korean manufacturer Carima, you will notice that the print settings associated with these printers have a very limited set of controls. As shown in *Figure 9.21*, the only print setting you can modify when editing a print setting related to a Carima 3D printer is **Layer Thickness**. This is because all the key controls have been set in Fusion 360 as provided by the printer manufacturer. This means that if you need further customization for these print settings, you will need to contact Carima support, and they will be able to provide a file containing custom parameter values, which you can simply load into your custom print setting in Fusion 360.

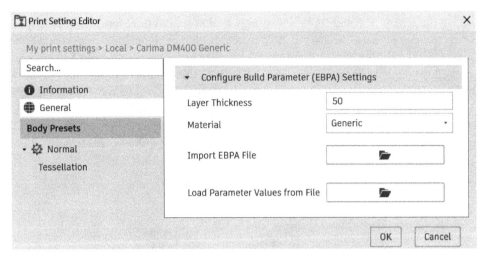

Figure 9.21 – Carima DM 400 Generic print settings

If you have an SLA 3D printer from Prusa Research such as SL1S SPEED, you will see additional inputs that you can modify, such as **Exposure Time**, **Initial Exposure Time**, and **Faded Layers**. The print settings included in Fusion 360's **Print Setting Library** for this specific printer are shown in *Figure 9.22*. The figure shows the **General** print settings for the **Prusa Orange Tough** resin. If you are using a different resin, you may need to modify the **Exposure Time** inputs and customize your print settings accordingly so that you have sufficient cure times for your resin during printing.

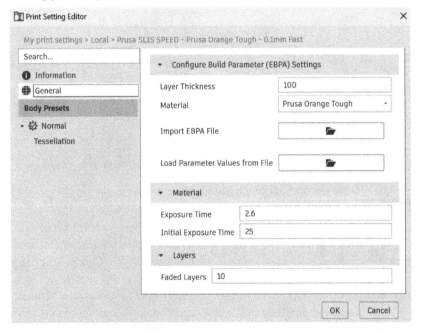

Figure 9.22 – Prusa SL1S SPEED, the Prusa Orange Tough print setting

In this section, we covered the most important inputs for print settings for SLA and DLP 3D printers. We highlighted how you can customize a print setting by saving it to your local library. We demonstrated how **Layer Thickness** can be edited regardless of the print setting type. We highlighted the differences between print settings for a generic SLA and DLP printer versus machine vendor-specific print settings from Anycubic, Carima, and Prusa. In the next section, we will focus on how to assign a unique print setting to portions of our design to achieve specific manufacturing criteria.

Assigning unique print settings to bodies

In the previous sections, we covered how to select and modify print settings for FFF and SLA/DLP 3D printers. We highlighted how each print setting contains multiple body presets within it and how you can modify the information within each body preset to customize the print setting for your needs. For FFF printers, the print settings that are shipped with Fusion 360's library include a preset for **Normal** and **Strong** part printing options. We talked about how to change the default preset from **Normal** to a secondary preset such as **Strong** or a custom one we can create by duplicating an existing one. In this section, we will talk about how to assign those presets to specific bodies within our additive setup.

To demonstrate this workflow, we will create a second setup using a second manufacturing model, as shown in *Figure 9.23*. As you can see in the figure, this setup is for an FFF printer (**Prusa i3 MK3S +**) and utilizes the **ABS (Direct Drive)** print setting. This setup includes all three components that were present in the original design document. They have been automatically positioned along the *Y* axis using **Additive Arrange** with an **Object Spacing** input of 5 millimeters.

By default, all three components will utilize the **Normal** body preset, as this is the default preset for the **ABS (Direct Drive)** print setting. However, we can easily assign a different body preset to the body of our choosing. If we select the **Additive Arrange** line item from the browser and right-click on it, we can choose the **Assign Body Presets** command. This will bring up a new dialog called **ASSIGN BODY PRESETS** and display the available body presets within the selected print setting. We can see both the **Normal** body preset and the **Strong** body preset within this dialog. The **Strong** body preset row is highlighted in blue, waiting for us to select a body to assign it to.

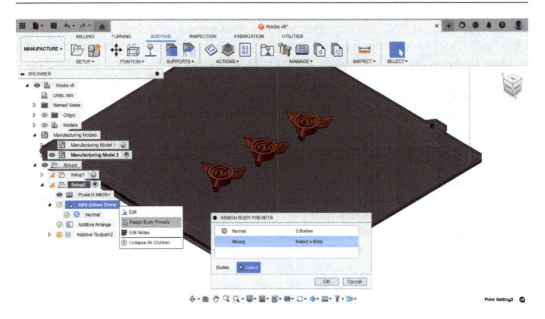

Figure 9.23 – The ASSIGN BODY PRESETS dialog in an additive setup

In order to assign the **Strong** body preset to a given object, all we have to do is select the body. After selecting the body in the middle of the build volume, as shown in *Figure 9.24*, the **ASSIGN BODY PRESETS** dialog will update automatically, and we will see that the **Normal** body preset is assigned to two bodies and the **Strong** body preset is assigned to the one body we have just selected.

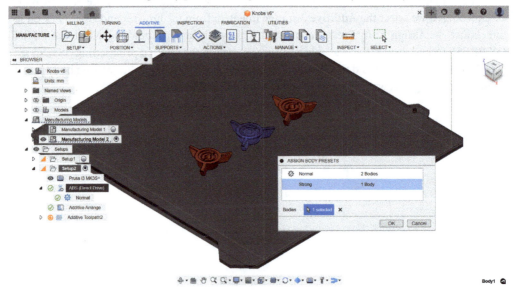

Figure 9.24 – Assigning a non-default body preset to a selected body

After selecting the **OK** button within the **ASSIGN BODY PRESETS** dialog, the browser will automatically update and display both the **Normal** body preset and the **Strong** body preset under the **ABS (Direct Drive)** folder for the active setup, as shown in *Figure 9.25*. This indicates that the active setup has bodies utilizing multiple body presets. To demonstrate the impact of different body presets on different bodies, we can select the **Additive Toolpath** line item from the browser and generate the toolpath using the *Ctrl + G* keyboard shortcut. Next, we can activate the **Simulate Additive Toolpath** command located within the **ACTIONS** panel within the **ADDITIVE** tab:

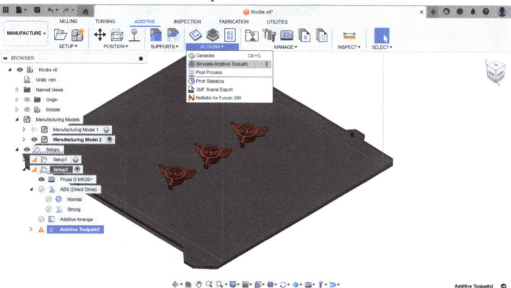

Figure 9.25 – Simulating the additive toolpath for an additive setup

When we are visualizing the additive toolpath, we can zoom in on our model so that we can see the two neighboring components a little closer, as shown in *Figure 9.26*. If we uncheck the **Show Machine** checkbox, the build plate will be hidden so that we can easily differentiate the colors of the toolpath type.

In the same figure, you will notice that the knob on the bottom of the image only has a single red line (outer perimeter) and a single green line (inner perimeter). The knob on the top has a single outer perimeter colored in red but has four green inner perimeter toolpath lines for the layer currently being displayed (layer 41). This is because the **ABS (Direct Drive)** print setting has an input of 5 within the **Number of Perimeters** field located in the **Shell** subheading of the **Strong** body preset.

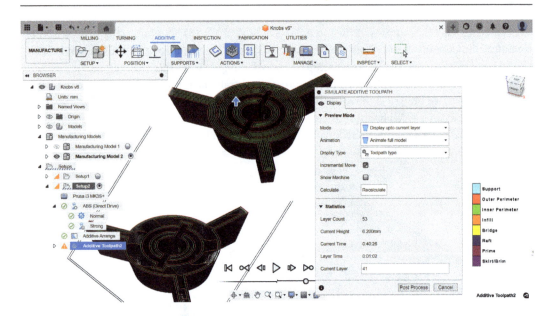

Figure 9.26 – Simulating the additive toolpath for an FFF setup with two different body presets

There are many reasons we may want to assign a different body preset to selected bodies. In the previous example, we demonstrated how you can print different bodies with different settings for toolpaths, such as the number of perimeters to make one of them stronger than the other. Another reason we may want to utilize a body preset and assign it to a given body is to control which extruder is used to print a specific body.

In the next example, we have a single component that is made up of two bodies. The component is named **2BC** and the bodies are named **Body1-Yellow** and **Body2-Red**, as you can see in *Figure 9.27*. If we create an additive setup with a 3D printer that has multiple extruders, such as **Raise 3D Pro 2**, we can assign a print setting and later control which body is printed with which extruder. In this example, we will assign **PLA (Bowden Tube)** as the print setting after selecting our printer. Next, we can select the print setting and right-click on it to edit it. This will also allow us to customize the name of the body presets. *Figure 9.27* demonstrates how we can change the name of the default body preset from **Normal** to `Yellow`.

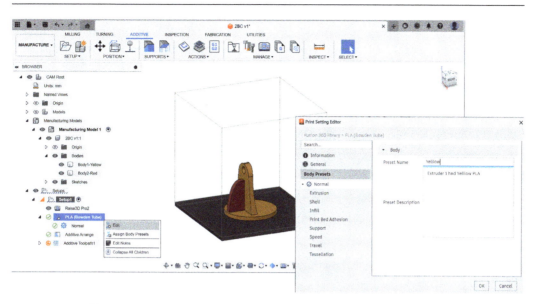

Figure 9.27 – Customizing the name of body presets within a print setting

After renaming the existing body presets to Yellow and Red, respectively, we will have to change which extruder each body preset relies on. As you can see in *Figure 9.28*, we can go to the **Extrusion** subheading and change **Part Extruder** from **Extruder 1** to **Extruder 2** for the **Red** body preset.

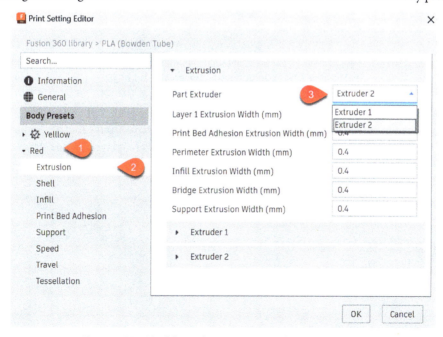

Figure 9.28 – Modifying the part extruder for a body preset

After assigning a unique extruder to each body preset, we can select **OK** in the **Print Setting Editor** dialog to accept our changes. Next, we will have to assign a body preset to a given body. We can do that by right-clicking on the **PLA (Bowden Tube)** print setting within the browser and selecting **Assign Body Presets**, as shown in *Figure 9.29*. Next, we can select the body we wish to print using **Extruder 2**. The **ASSIGN BODY PRESETS** dialog now shows that one body will be printed using the body preset named **Yellow** and the second body will be printed using the body preset named **Red**. After clicking **OK**, our body preset assignment is now complete.

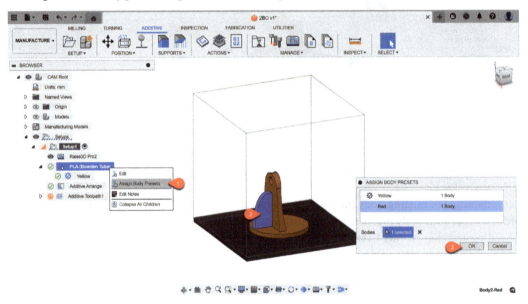

Figure 9.29 – Assigning body presets to a body

As you can see in *Figure 9.30*, our browser updates and we can see both **Yellow** and **Red** as body presets that are active in this setup. After generating the additive toolpath and simulating it just as we have done in the previous example, we can toggle through the different layers of our sliced model. *Figure 9.30* shows a snapshot from layer 310.

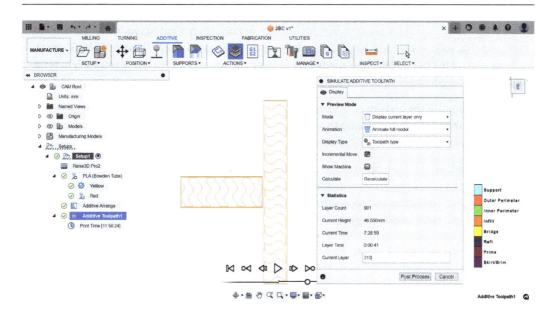

Figure 9.30 – Simulation of the additive toolpath

The preceding figure shows the slice 46.55 millimeters above the build platform. The **Show Machine** checkbox has been deselected within **SIMULATE ADDITIVE TOOLPATH** to hide the build plate. The view is set to **TOP** from the View cube. **Preview Mode** is set to **Display current layer only**. This combination of settings shows us the toolpath for the current layer and hides everything else. The horizontal box that we see in the preceding figure represents the body that will be printed using the red filament with **Extruder 2**. The vertical box that we see in the same figure represents the body that will be printed using the yellow filament with **Extruder 1**. You will notice that these two boxes have a distinct perimeter around them as they are two different bodies that are being printed with two different extruders. This means that if you print this toolpath, the interface between these two boxes will be more pronounced versus if you were to print them using the same body preset or as a single joined body with a uniform perimeter and a continuous infill.

Summary

In this chapter, we focused on print settings, including how to customize them and how to assign unique body presets to given parts. This was a long chapter so let's do a quick recap of what we learned.

Fusion 360 has a library of print settings you can use when creating your additive setups. Each print setting has a set of default values that you can use for printing with your specific printer and material combination. However, there will be instances where you may want to modify some of these settings to either improve your print quality or fit the needs of your specific printer. In such cases, you can select the print setting you wish to modify from Fusion 360's library and copy it over to your personal Print Setting Library folders. Once you have a print setting in your local library, you can modify it and save it for future use. There are certain print settings you definitely want to modify based on your filament or resin, such as the layer height, the temperature with which you will extrude the filament, the temperature of the build plate, how many perimeters and top and bottom layers you wish to print to control the strength of your print, and your print bed adhesion settings.

You can also customize the body presets to your liking and later assign a unique body preset for each body within your additive setup. Using these techniques, you will have full control over the speed and the quality of your 3D prints.

In the next chapter, we will learn how to create support structures for parts and additive manufacturing technologies that require support. We will also get a better understanding of how Fusion 360 interprets those supports during slicing and toolpath generation based on the print settings for the support structures we highlighted in this chapter.

10
Support Structures

Welcome to *Chapter 10*. In this chapter, we will learn all about how to generate and slice support structures for **fused filament fabrication (FFF)**, **stereolithography (SLA)**, and **digital light processing (DLP)** 3D printing.

Most mainstream additive manufacturing technologies (except for binder jetting, selective laser sintering, and multi-axis deposition) require support structures to be generated and printed for sections of components that have large overhanging surfaces. In this chapter, we will highlight the bar and volume support structures, which can be used for both FFF and SLA/DLP 3D printing.

Almost everything we will learn in this chapter also applies to generating support structures for parts to be printed with **metal powder bed fusion (MPBF)** printers, but we will dive deeper into positioning parts and generating support structures for the MPBF process in *Chapter 12*. As the MPBF technology is capable of printing parts and support structures with higher precision compared to FFF, it is not recommended to use the solid FFF supports such as volume and bar structures in an MPBF setup. The MPBF process generally utilizes hatch-style volume supports, which is why we have a dedicated chapter on how to support parts for this process.

In this chapter, we will highlight how the support structures we generate are sliced based on the printing technology used in our additive setup. We will also learn about several unique support connections that can be added to the build plate, such as pads, roots, and a base plate, which are commonly used to connect bar supports to the build plate more effectively and increase build plate adhesion when 3D printing with the bottom-up SLA/DLP technologies.

In this chapter, we will cover the following topics:

- Volume and bar supports for FFF printing
- Bar supports for SLA and DLP printing
- Base plate supports for SLA and DLP printing

By the end of this chapter, you will have a better understanding of how to generate support structures for both solid and mesh models using Fusion 360. You'll learn how to create bar-shaped supports including lattices, volume supports, and base plate supports depending on the 3D printing technology. You will learn the differences between top-down and bottom-up resin printing technologies and their unique support requirements. You will also gain a solid understanding of how to create bar supports over an area, an edge, or a local minima point in order to successfully 3D print parts using SLA/DLP printers.

Technical requirements

Everything covered in this chapter is accessible to users with the personal, trial, commercial, startup, or educational Fusion 360 license types.

There are certain Fusion 360 additive machines, such as **metal powder bed fusion** (**MPBF**) 3D printers and their respective print settings and support structure generation capabilities, which are not available until gaining access to the Manufacturing extension. As with all other Fusion 360 extensions, the Manufacturing extension is not available for personal use. As we will not cover topics specific to metal 3D printing in this chapter, you should be able to follow along with all the examples in this chapter with no issues.

If you have a commercial license of Fusion 360, to create an additive setup with a metal powder bed fusion machine, you will also need access to the Manufacturing extension (`https://www.autodesk.com/products/Fusion-360/additive-build-extension`).

Autodesk offers a 14-day trial for the Manufacturing extension for all commercial users of Fusion 360. You can activate your trial within the extensions dialog of Fusion 360.

The lesson files for this chapter can be found here: `https://github.com/PacktPublishing/3D-Printing-with-Fusion-360`.

Volume and bar supports for FFF printing

Fused filament fabrication (**FFF**) 3D printers can generally print parts with overhang angles of less than 45 degrees without needing support structures. If you want to print a part using an FFF printer, one of the first things to do is to try and find an orientation that will allow you to print that part with no support structures.

However, certain models cannot be printed without supports. This is because no matter which orientation you choose, certain parts will end up with down-skin surfaces with overhangs larger than 45 degrees that span distances larger than you can bridge with your 3D printer. In such cases, you will need to generate support structures using your CAD tool or your slicer. As Fusion 360 is both a CAD software and a slicer, you can generate support structures parametrically within the MANUFACTURE workspace for any additive setup.

In this section, we will talk about how to generate volume and bar supports for an additive setup with an FFF 3D printer. To demonstrate how to generate volume supports, we will use a Fusion 360 document named `Bracket.f3d`. This document already contains an additive setup within the **MANUFACTURE** workspace. As shown in *Figure 10.1*, **Setup 1** is an additive setup with a **Prusa i3MK3S** + printer and a **PLA (Direct drive)** print setting. The component has been arranged within the 3D printer such that it is in the middle of the build volume. A suitable orientation has also been selected for this part. Based on this orientation, we will need to generate support structures in order to print this part successfully. *Figure 10.1* shows the two options we have to generate support structures located within the **SUPPORTS** panel of the **ADDITIVE** tab for this setup.

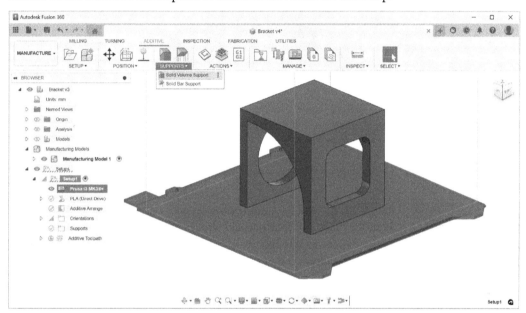

Figure 10.1 – Support structure types for an additive setup with an FFF 3D printer

If you select the **Solid Volume Support** command located within the **SUPPORTS** panel, a new dialog called **SOLID VOLUME SUPPORT** will become visible. In this dialog, there are two tabs. The **Geometry** tab allows you to select which faces or bodies should be supported with this operation. *Figure 10.2* shows how the selected faces are highlighted in blue while the **SOLID VOLUME SUPPORT** dialog is active.

Within this dialog, you also have the ability to control the Support Overhang Angle. Fusion 360 will automatically generate a support structure for sections of downward-facing surfaces that have an overhang angle smaller than the entry in this input field. In other words, if we increase the Support Overhang Angle input, Fusion 360 will generate supports over a larger area.

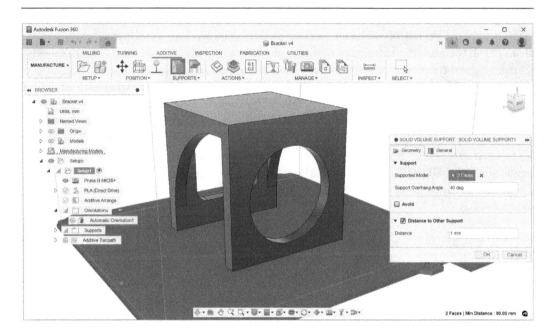

Figure 10.2 – Selecting faces of a solid model for a volume support

In the GitHub folder for *Chapter 10*, you will find another file called `Bracket-Mesh.f3d`. This Fusion 360 document is a duplicate of the previous one with one small difference. As you can see in *Figure 10.3*, in this second document, the component named **Bracket** is not a solid body, but a mesh body. The solid volume support dialog allows you to select not only entire bodies but also individual face groups when the component being supported is a mesh body. In order to select individual face groups, you have to make sure that the mesh face groups are displayed. The easiest way to display mesh face groups while you are within the **MANUFACTURE** workspace of Fusion 360 is to use the keyboard shortcut *Shift + F*. If the mesh body in the active setup has more than one face group, each mesh face group will be displayed with unique colors, as you can see in *Figure 10.3*. After displaying the mesh face groups within the **MANUFACTURE** workspace for a given additive setup, you can select face groups as input objects to be supported within the **SOLID VOLUME SUPPORT** dialog.

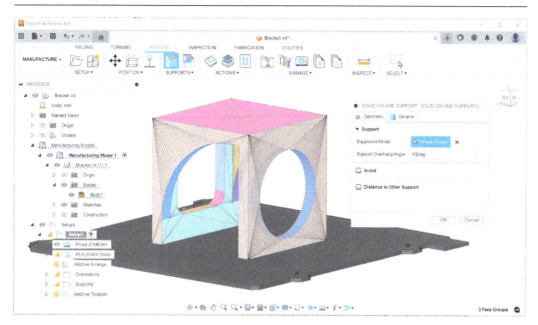

Figure 10.3 – Selecting three face groups of a mesh body

Now that we have highlighted how to select face groups for a mesh body when generating volume supports, we can switch back to the original model (`Bracket.f3d`) and create our volume supports on the previously selected two faces of the solid body. Once the support structure is generated by clicking the **OK** button on the SOLID VOLUME SUPPORT dialog, Fusion 360 will create a support structure as shown in *Figure 10.4*, and you will be able to visualize it as a blue volume under the overhanging surfaces. Fusion 360 also adds a new line item called **Solid Volume Support1** within the `Supports` folder of the active setup in the browser.

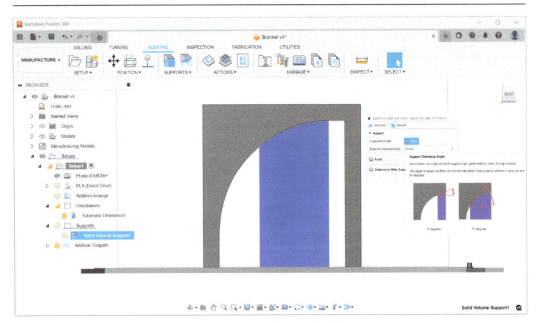

Figure 10.4 – Visualizing a solid volume support

In *Figure 10.4*, you may have noticed that the blue-colored support structure does not enclose the entire opening, but rather supports roughly half the arc face. This is because we used a support overhang angle of 40 degrees and Fusion 360 did not support any point on that arc with an overhang angle greater than 40 degrees. It is important to remember that overhang angles are measured from the horizontal plane as demonstrated by the tooltip image in the preceding figure.

Another option we have within the **Geometry** tab of the **SOLID VOLUME SUPPORT** dialog is the **Avoid** checkbox. By activating this checkbox, we can select one or more faces as input and prevent Fusion 360 from generating support structures on those surfaces. *Figure 10.5* shows the creation of a second Solid Volume Support to support the entire body, while avoiding the generation of supports on the two arc faces highlighted in blue, which were previously used as inputs to create **Solid Volume Support1**.

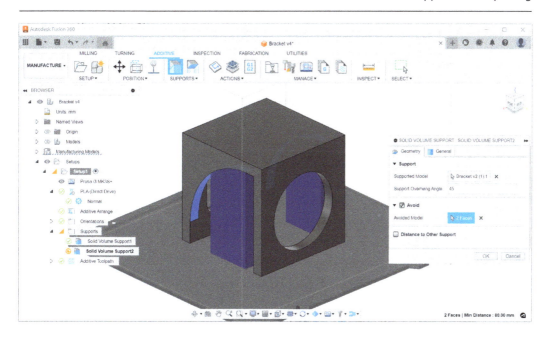

Figure 10.5 – Selecting faces to avoid while generating a solid volume support

The outcome of the generation of this second solid volume support is the creation of a volume support for the entire body, while excluding the two faces we specified in the dialog to avoid. Using this technique, we can add unique support structures to certain faces while creating a different set of supports for the rest of the body.

The second tab within the **SOLID VOLUME SUPPORTS** dialog is named **General**. *Figure 10.6* shows the two input fields in the **General** tab. The first input field, named **Top Distance to Part,** controls the distance between the support structure we are generating and the down-skin of the part. The second input field, named **Bottom Distance to Part,** controls the distance between the support structure and the up-skin of the part if a given support structure were to be touching the up-skin of our part instead of the build plate.

The inputs for both of these fields are in units of length, and by default, the input is 0.3 millimeters. As shown in *Figure 10.6,* the layer height for the PLA (Direct Drive) print setting is 0.15 millimeters. Therefore, a support structure we generate using these settings will have a two-layer gap above the up-skin and below the down-skin of our part.

Based on my experience, when 3D printing with PLA, whenever we leave a two-layer gap between the part and the supports, the post-processing becomes easier, and we can remove the support structure from the part after printing without a lot of effort required.

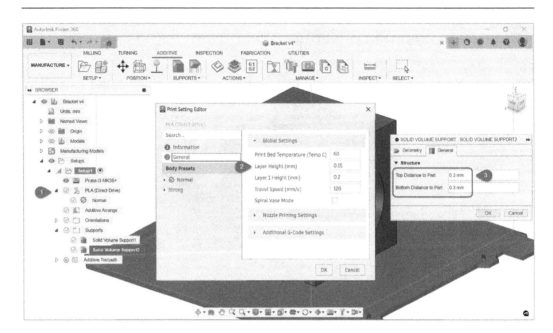

Figure 10.6 – The relationship between volume supports and print settings

After generating one or multiple support structures, Fusion 360 adds them to the supports folder within the browser, as shown in *Figure 10.7*. If we select one of those support structures, we can right-click on them and see a menu of options. We have options to **Suppress** a given support, **Duplicate**, **Cut**, **Copy/Paste**, or **Delete** it. Keep in mind that the order of support structures within the supports folder is important. Subsequent supports are aware of the existence and location of their predecessors and can avoid generating supports where one has already been created. Therefore, if you were to suppress a given support within the supports folder, a preceding support structure may become out of date and may need to be regenerated.

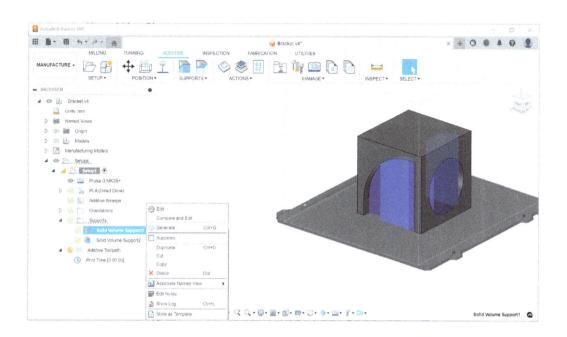

Figure 10.7 – Right-click menu options for support structures

Now that we have learned how to generate volume supports for an FFF setup, we can focus on generating bar supports. *Figure 10.8* shows the **SOLID BAR SUPPORT** dialog, which can be initiated by the command with the same name located within the **SUPPORTS** panel in the **ADDITIVE** tab.

Within the **Geometry** tab of this dialog, the first thing you can do is select a set of faces or bodies to be supported. As you may have noticed, the options in the **Geometry** tab for bar supports and volume supports are exactly the same. In this tab, you can enter the support overhang angle, select which surfaces you wish to avoid creating support structures for, and designate a gap between multiple support entities to avoid overlap between support structures.

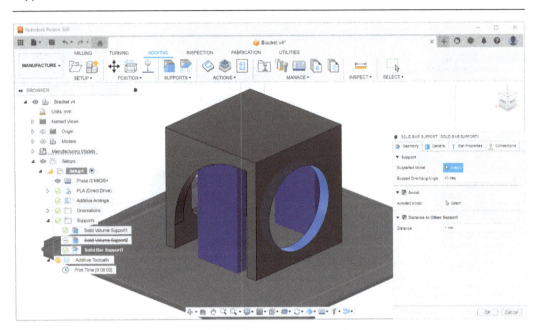

Figure 10.8 – Selecting faces for generating bar supports

The options to control the density of the connection points between the part and the support structure are located in the **General** tab, as shown in *Figure 10.9* and labeled **1**. Fusion 360 refers to these connection points as anchors. Within the **General** tab, you also have the option to group multiple bars and bunch them up into a tree/bouquet structure, and control the shape of those structures, labeled **2** in the same figure.

The term *anchor* within the **Anchor Density** field of the **General** tab refers to connection points between the bar support and the down-skin of a given part. Fusion 360 offers several preset anchor densities such as **Sparse**, **Medium**, and **Dense**, as well as the ability to fully customize the anchor density, which allows you to control the distance between the anchor points, their alignment on the downskin face, and the distance between the anchor points and the perimeter of the downskin face. Using a custom anchor density, you can also distribute the anchor points in a hexagonal or rectangular pattern.

Similar to how you can control the anchor placement settings, Fusion 360 also allows you to control the inputs for the bouquet/tree structures for bar supports. You can choose from a set of presets including **Small**, **Medium**, and **Large** bouquet/tree structures while generating bar supports. Alternatively, you could choose to eliminate the bouquet/tree structures altogether by selecting **None** from the drop-down list. If you like, you can also fully customize the bouquet/tree structure, as shown in *Figure 10.9* and labeled **2**. After selecting **Custom** from the drop-down menu for bouquet/tree type, you will have access to additional controls such as the diameter and height of a given bouquet/tree structure as well as its recursive depth. You can also control the maximum vertical angle at which your bouquets can branch off from the main column of each bar support.

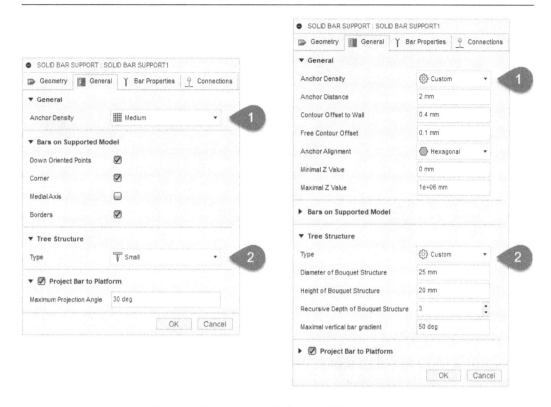

Figure 10.9 – The General tab for solid bar supports

The **Bar** Properties tab allows you to control the size and shape of your bars, as shown in *Figure 10.10*. The **Bar Size** section allows you to control the diameter of your bars. If the **Width on Part** and **Width on Platform** inputs are the same, you will have a bar with a uniform cross-section. You can also generate a bar that resembles the frustum of a cone if you enter a larger value in the **Width on Platform** input versus the **Width on Part** input.

The **Bar Shape** section of the solid bar support dialog allows you to customize the shape of your bars. The default bars generated by Fusion 360 have a **Polygon Corner Count** value of **5**. This results in bar supports with a pentagonal cross section. If you increased the **Polygon Corner Count** input to **6** or **8**, you would end up with hexagonal or octagonal cross sections, respectively. You could also generate elliptical cross sections by changing the **Radial Type** dropdown from **Circular** to **Ellipse** and increasing the **Ellipse Stretch in** % input field from **100%** up to **200%**.

The **Bar Path** section has additional settings that allow you to modify the route a bar support follows on its way to the build plate. As shown in *Figure 10.10*, bar supports are connected to parts at a right angle by default. Just like volume supports, the default bar support settings in FFF setups also leave a gap between the part and support structure. The default distance for this gap is 0.15 millimeters as set in the **Distance to Part** input field. If you are using a print setting of 0.15 mm, this gap equates to a 1-layer gap between the bar supports and the down-skin of your parts.

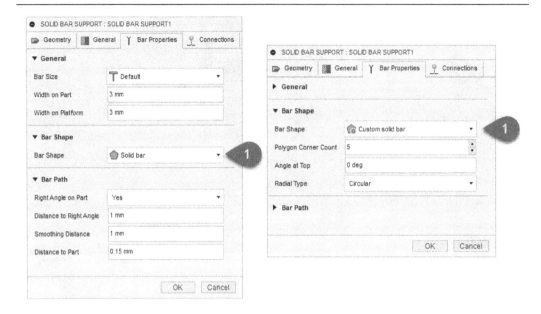

Figure 10.10 – Selecting a custom bar shape for the solid bar supports

The **Connections** tab of the **SOLID BAR SUPPORT** dialog allows you to control the shape of the breaking point between the bar support and the part. The default settings for bar supports in an FFF setup do not use a break point, as indicated by **1** on the left in *Figure 10.11*. If you choose the **Custom** option from the drop-down menu as indicated by **1** on the right within the same figure, you can control the size and shape of the break point using the additional input fields that become visible.

You also have the ability to control how the bar supports are attached to the build plate. By default, all bar supports have a solid connection to the build plate as the **Pad on Platform** checkbox is active, as shown in *Figure 10.11*. You can control the diameter of the pad as well as its height and taper angle using the relevant input fields.

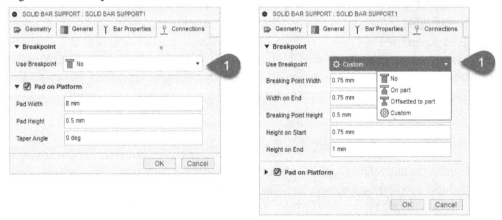

Figure 10.11 – Selecting a custom breakpoint for solid bar supports

Figure 10.12 shows the outcome of generating bar supports on the circular face for this part. To generate this specific support, you would need to use the following settings:

- Bar Properties tab - Bar Size = Default

- Bar Properties tab - Bar Shape = Solid Bar

- Bar Properties tab - Right Angle on Part = Yes

- Bar Properties tab - Distance to Part = 0.3 mm

- Connections tab - Use Breakpoint = No

- Connections tab – Pad on Platform = Unchecked

As shown in the preceding list, we increase the input for **Distance to Part** to 0 . 3 millimeters so that we get a two-layer separation between the part and the support. In *Figure 10.12,* you can visualize the gap between the bar supports and the part as indicated by the empty white space between the blue-colored bars and the gray-colored part. We also unchecked the **Pad on Platform** option when generating this support structure so that the bars touch the build plate without occupying any additional area on the build plate.

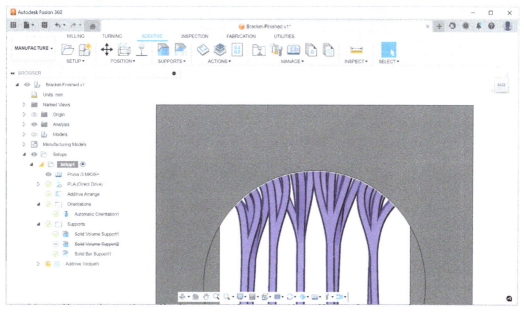

Figure 10.12 – Bar supports with a bouquet structure

If we zoom out from the section where the bar supports touch the part, we can also see how this bar support connects to the build plate. *Figure 10.13* shows the same part and its bar support using a **Section Analysis** view where the part and the support structure have been sectioned off using the YZ plane. In the following figure, both of the Solid Volume supports have been suppressed from the browser to allow us to view the bar supports without any obstruction. You will notice that this bar support utilizes a small tree/bouquet structure and the bars have been projected to the platform with a **Maximum Projection Angle** value of 3 0 degrees so that the bar support path is toward the build plate while avoiding any up-skin on the part itself, which will make the post-processing step of removing supports easier.

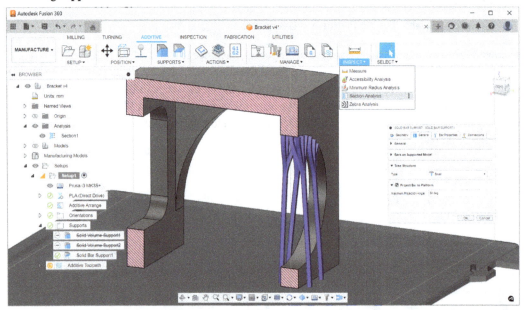

Figure 10.13 – Bar supports projected to the platform

To get a better understanding of how bar and solid support structures will be printed, we need to generate the additive toolpath and simulate it. To get started, let's unsuppress **Solid Volume Support1**. We will keep **Solid Volume Support2** suppressed as shown in *Figure 10.14*.

Fusion 360 has the ability to create additive toolpaths in a unique way for volume supports versus bar supports. As you may recall from *Chapter 9*, the print setting editor controls how FFF supports are sliced and how additive toolpaths are generated. You can control options such as how many perimeters are created for each support type.

In this exercise, we haven't customized our body preset and are utilizing the **Normal** body preset for our component and its associated support structures. If we edit the print settings and observe the **Support** subsection for the **Normal** body preset, we can see that Volume supports have a value of 0 for Number of Support Perimeters, whereas bar supports are sliced using a perimeter of 1 when the toolpaths are generated.

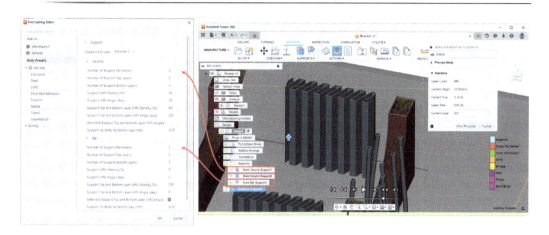

Figure 10.14 – The toolpath visualization for bar and volume supports in our FFF setup

After generating the additive toolpath and simulating it, we can see that the light blue colored object in the additive simulation represents the support structures. *Figure 10.14* shows the additive toolpath at layer 300, which happens to be at a height of 45.05 millimeters as this part is sliced at a layer height of 0.15 millimeters. If you look at the region where the volume supports are, you will notice that they are sliced using an accordion-style toolpath. This is due to the 15% infill density setting for **Volume supports** in the print setting editor for the **Normal** preset. You can also observe in the same figure that the toolpath for the volume support does not contain a perimeter line surrounding the support.

In contrast to volume supports, the toolpath generated for bar supports has a pentagonal cross-section with a continuous line around its perimeter. However, the bar support toolpath does not contain any infill, as the support infill density for bar supports is set to 0% for the **Normal** preset in this same print setting.

This example demonstrated how we can generate solid volume supports and bar supports to support our parts when printed using FFF 3D printers. We learned how to control the various settings to generate the desired shapes and connection types for bar and volume supports. We also learned how Fusion 360 generates the slices and toolpaths for these support structures. Using these settings, we can create any support structure we need for a successful print. In the next section, we will learn how to create various bar supports to manufacture parts using SLA and DLP 3D printers.

Bar supports for SLA and DLP printing

There are two types of resin-based printers on the market: bottom-up resin printers and top-down resin printers. In a bottom-up resin printer, the light source is below the resin tank. As the build plate lifts up from the resin tank (sometimes referred to as the *vat*), the part is printed upside down. In a top-down resin printer, the light source is above the build plate. As the build plate is submerged down into the resin tank with each layer, the part is printed right-side up. Both types of resin 3D printers require support structures for different purposes.

In top-down resin 3D printing, we need to generate support structures for overhanging surfaces similar to how we generated them for FFF 3D printing. However, the support structures we will generate for resin printers will most likely be bar supports instead of volume supports.

In top-down 3D printing, with each layer, the build plate is submerged further into a resin tank. As there is no **fluorinated ethylene propylene (FEP)** film between the layer being printed and the resin, this printing process is not subjected to any separation force between the part and the FEP film. Therefore, we can generally create weaker support structures and still manufacture our parts successfully.

Bottom-up resin printers need to overcome gravity as they lift the part out of the resin vat. Bottom-up resin printers also need to overcome the peeling effect during the printing process between the part and the FEP film. Depending on your part shape and your drainage hole positioning, you may also have to overcome the cupping effect when using a bottom-up resin printer. This is why the bottom-up resin printing process generally requires stronger support structures.

Fusion 360 has four distinct commands for adding bar supports to additive setups for SLA /DLP 3D printers. In *Figure 10.15*, you can see an additive setup with a single component. The Fusion 360 design named Hook.f3d is located in the GitHub folder for this chapter. This file contains an additive setup named **Setup1** for a Formlabs Form 3 printer and its associated print settings. The component has been automatically oriented and is above the build plate by 7 millimeters, as shown in the following figure. When adding bar supports to an SLA/DLP additive setup, the **BAR SUPPORT** dialog consists of four tabs. *Figure 10.15* shows the **Geometry** tab for bar supports. As you can see in the screenshot, after selecting the model to be supported, we can control the **Support Overhang Angle** input.

Just like the solid volume support and bar support dialogs we explored in the previous section of this chapter, the bar supports for SLA/DLP printing measure the overhang angle from the horizontal plane (XY plane). Fusion 360 generates supports structures for overhanging surfaces with an angle value below that specified in the **Support Overhang Angle** input field.

You may have also noticed that the **Geometry** tab for SLA/DLP bar supports is similar to the **Geometry** tab for volume and bar supports for FFF additive setups seen previously. In addition to the **Support** section, we also have the **Avoid** checkbox and the **Distance to Other Support** checkbox that allow us to specify which surfaces not to support and how far away from a neighboring support structure this bar support needs to be.

There is one additional checkbox in the **Geometry** tab named **Advanced Area Filter**. When activated, the advanced area filter allows us to control the minimum and maximum area to be supported and whether we would like Fusion 360 to generate supports for faces where other support structures may already exist. We can also control the minimum and maximum Z heights of the bar supports to be generated. By default, these entries are 0 millimeters and 10106 millimeters, respectively. If we were to reduce the **Maximal Z Height** input to 50 mm, Fusion 360 would only create bar supports for sections of faces with overhang angles of less than 45 degrees located below the first 50 millimeters of our part.

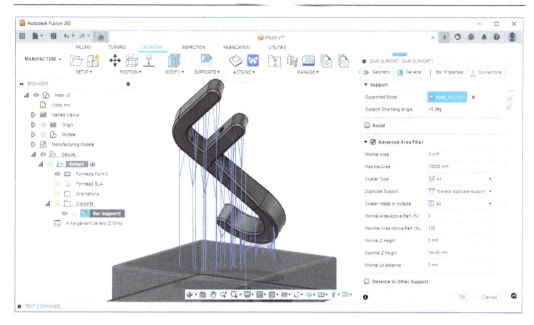

Figure 10.15 – Geometry tab for bar supports on SLA/DLP printers

By default, Fusion 360 creates bar supports utilizing a bouquet/tree structure as shown in the previous figure. We can modify the type of bar support Fusion 360 generates, and how close those bars should be from each other while touching the part, within the **General** tab of the **BAR SUPPORT** dialog, as shown in *Figure 10.16*. By changing the **Anchor Density** input from the default option of **Medium** to **Dense**, we can add more touch points between the bar supports and the part. We can also change the shape of the tree/bouquet structure. The following figure shows the effect of turning off the tree structure by selecting **None** from the **Type** drop-down menu.

Fusion 360 also allows you to group bars together by activating the **Groups** checkbox. *Figure 10.16* shows 5 bars grouped together with a **Group Taper** (%) input of 5 0 percent.

This means that approximately five bars on their way from the part to the build plate bunch up together toward a central point by the specified amount. If you print your part and supports with this setting, you can handle groups of five bars at a time and break them off the part during post-processing much easier.

In the **General** tab, you can also turn on the **Brace** option for bar supports. *Figure 10.16* shows bars with **K** type bracing. Fusion 360 offers three different brace types that resemble the letters K, X and N. Using the additional options below the **Brace** checkbox, you can control how many bars you wish to link together, what the link angle should be between the brace and the main column of your bar support, and whether you want the connection between the brace and the main column of the bar support to be straight or curved.

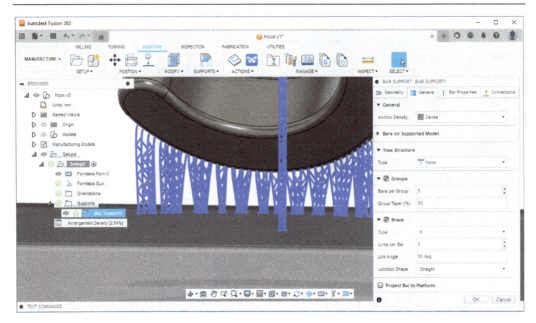

Figure 10.16 – Grouped and braced bar supports with no tree structure

The **Bar Properties** tab for the bar supports in SLA/DLP printing is the same as that for FFF 3D printing. Within the **Bar Properties** tab, you can control the bar size and the bar shape. The **Bar Properties** tab also allows you to control the path your bars should follow between the part and the platform. You can either follow a path with a right angle to the down-skin surface normal, or your bars could point straight down to the build plate.

Now let's move over to the final tab named **Connections**. In this tab, you can control how the bar support is connected to the part as well as the platform. By default, bar supports are connected to the parts using a breakpoint on the part. You can also choose to have your breakpoint offset from the part or have no breakpoint at all. You can also customize the size and shape of your breakpoint using exact specifications for how the bar support should connect to your part.

The next checkbox named **Pad on Platform** allows you to add a small circular puck that is slightly larger than the bar itself where the bar support connects the build plate. If you turn this checkbox on, you can then control the pad's width, height, and taper angle. A taper angle of zero will give you a pad with a cylindrical cross-section. A positive taper angle will result in a frustum shape where the pad is bigger at the platform and gets smaller as it approaches the bar. A negative taper angle will give you the opposite shape where the pad's connection to the build plate has a smaller cross-section and the pad gets wider as you move up in the Z direction. Pads with negative taper angles allow you to separate the pad and bar support from the build plate using a chisel with ease.

The last section within the **Connections** tab is named **Connection Type**. Fusion 360 offers three options within the **Type** dropdown: **None**, **Base plate for groups**, and **Roots**. *Figure 10.17* shows bar supports generated with the **Base plate for groups** option enabled. When this option is selected, you can create a base plate for the bars that you grouped together in the **General** tab. You can control the height of the base plate, as well as its offset from the perimeter of the group. You can also add a taper angle to these base plates as discussed previously while adding pads on the platform. Finally, you have the option to control whether to fill this base plate with a solid pattern or apply a hole pattern. We will cover pattern types in more detail when we discuss base plate supports in the next section.

If you choose **Roots** for the connection type, Fusion 360 will create bar support columns that separate into multiple roots just above the build plate. This creates a bouquet structure toward the build plate, creating a stronger connection between the trunk of the bar support and build plate and adding extra stability to tall and thin bar supports.

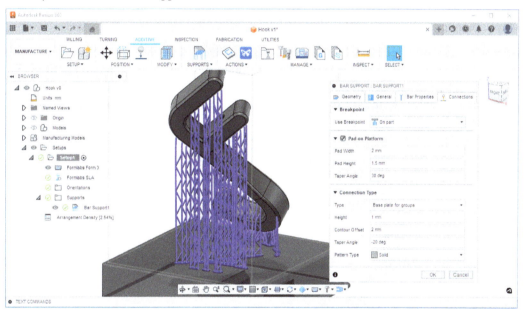

Figure 10.17 – Bar supports with pads and base plates for groups

Up until now, we have examined the Bar Support commands for SLA/DLP 3D printers. Fusion 360 can generate bar supports for entire components, as well as specific selected surfaces on a given component.

There may be times when you need to support down-oriented points and edges using bar supports. Fusion 360 has explicit commands to generate bar supports attached to down-oriented points and edges. The options to control the size and shape of the bars that are created using **Down-Oriented Point Bar Support** and **Edge with Bar Support** are exactly the same as the options we've already reviewed for bar supports. The main difference between bar supports and down-oriented point bar supports and edge with bar supports is that the down-oriented point bar support generates a single bar at the downward-facing corners, edges, and local minima of the selected component, whereas the **Edge with Bar Support** option generates bar supports along a downward-facing edge of the selected body or component.

When generating support structures for SLA/DLP printers, my recommendation is to start by adding a down-oriented point bar support first. This action will make sure that you always support any local minima on the part where you will most certainly need support structures. Next, I would recommend adding either a bar support or lattice volume support to the entire component so that all the surfaces that require support structures can be appropriately accounted for.

Now that we have mentioned lattice volume supports, let's take a closer look at this support action. *Figure 10.18* shows how to access the **Lattice Volume Support** option from the **ADDITIVE** tab in the **SUPPORTS** panel, and its various settings to control the shape of the bars, their connections to the part and the build plate, the size of the lattice cells, and their dimensional change as they move from the build plate to the parts.

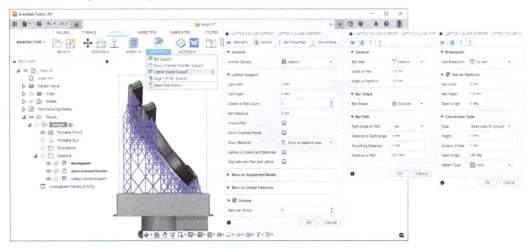

Figure 10.18 – Lattice Volume Support options for an additive setup

As we can see in *Figure 10.18*, there are a lot of similarities between the bar supports and lattice volume supports. Lattice volume supports are essentially bar supports with a unique bracing style. The main difference between lattice volume supports and bar supports is that you cannot select a certain brace type (K, X, or N) in a lattice volume support. Lattice volume supports have their own distinct bracing method. You can, however, control the shape and the growth ratio behavior of the lattice itself within the **General** tab under the **LATTICE VOLUME SUPPORT** section.

The preceding figure shows a lattice cell with a width of 3 millimeters and a height of 4 millimeters. It also shows how the lattice cell size changes and that you have larger cells at the base plate and smaller cells at the part connection zone. This change in cell size is controlled by the **Adaptive Step Count** input, which is 3 for this support structure.

You can also group bars together when generating lattice volume supports by activating the **Groups** checkbox in the **General** tab. If you activate **Groups** when creating a lattice volume support, you can add base plates for groups in the **Connections** tab for the lattice volume support as shown in *Figure 10.18*.

In this section, we have covered how to add various bar supports to components to be manufactured using SLA/DLP 3D printers. We discussed the similarities between bar supports, down-oriented point bar supports, and edge with bar supports. We also covered how to add lattice volume supports.

In the next section, we will focus on the specific method for adding base plate supports to our parts after generating bar supports for better build plate adhesion.

Base plate supports for SLA and DLP printing

When generating bar supports automatically using Fusion 360, Fusion adds pads to each bar support by default. Pads are added to bars where they connect to the build plate with a surface area larger than the bar diameter. This action ensures good adhesion between the support and the build plate. Pad connections are often recommended for printing small parts such as jewelry with an SLA/DLP printer. However, if you are printing a larger part with bar supports using a bottom-up resin printer, you should consider printing them using a base plate connection instead of pads between your bar supports and the build plate. This will ensure that you have a stronger build plate adhesion and increase your chances of printing success.

In the previous section, we covered how to group bars together and connect those groups to the build plate using base plates. Fusion 360 also has an explicit **Base Plate Support** action as can be seen in *Figure 10.19*. This action is located in the **ADDITIVE** tab, in the **SUPPORTS** panel drop-down menu.

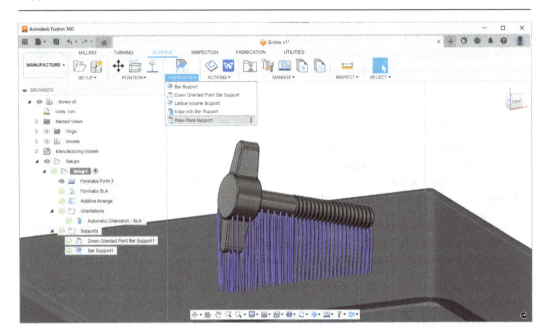

Figure 10.19 – How to access the Base Plate Support command in Fusion 360

Figure 10.19 shows an additive setup for a Formlabs Form 3 printer and its associated print settings. The part has been arranged and oriented deliberately. Two bar support operations have been applied to the part: **Down Oriented Point Bar Support1** and **Bar Support1**. This Fusion 360 document is named `Screw.f3d` and is included in the GitHub folder for this chapter.

After accessing the **Base Plate Support** command in the ribbon, you will see the **BASE PLATE SUPPORT** dialog as shown in *Figure 10.20*. This action also adds a new line item named **Base Plate Support1** to the **Supports** folder in the browser. After selecting the model to create the support within the **Geometry** tab, you can switch to the **General** tab. In the **General** tab, you will see the option to control the bounding shape of the base plate. You can either create a base plate based on the silhouette of the part based on its current orientation, or you can create a rectangular base plate based on the bounding box of the part. You can control the offset from the perimeter of the silhouette, the height of the base plate, and its taper angle. As we discussed in the previous section, a negative value used for the taper angle will create a base plate that increases in cross-sectional area in the positive Z direction. Such a shape allows you to use a chisel to separate the base plate support from the build plate with ease once the printing is complete.

In this example, we are supporting a solid part that does not have any holes. If we were printing a part with a hole, such as a ring being printed on its side, depending on which option we used for the **Fill Inner Contour** setting, we would generate different base plates. If the option was set to **Yes**, we would create a solid, circular, plate-shaped base plate. If the option was set to **No**, we would create a washer-shaped base plate.

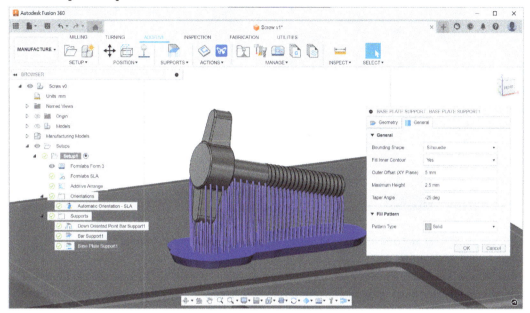

Figure 10.20 – Adding a base plate support in an additive setup for an SLA printer

In order to save material, you can also choose a pattern type within the **General** tab of the **BASE PLATE SUPPORT** dialog. The default pattern type is **Solid**. If you choose a different pattern type, such as **Hexagonal** as shown in *Figure 10.21*, you have the ability to control the size of the pattern as well as its wall thickness. Please be aware that if you are choosing a pattern other than solid, your bar supports may not always touch the printed segments of your base plate support depending on the size of your bars and the size of your pattern.

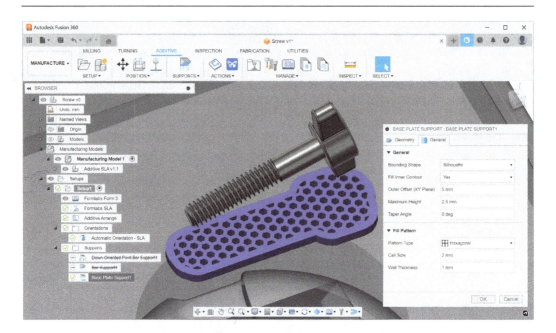

Figure 10.20 – Adding a hexagonal pattern for a base plate support

In this section, we covered how to add an explicit base plate support to increase the likelihood of print success for large parts. Base plate supports create a structure and a flat surface for the print to improve adherence to the build plate. By controlling the shape and thickness of the base plate support, you can ensure successful adhesion to the platform during 3D printing with an SLA/DLP 3D printer.

Summary

Whether you are planning on 3D printing using an FFF printer or a top-down or bottom-up SLA/DLP printer, you will need to generate support structures in order to successfully print overhanging areas or islands in your sliced model. Depending on your 3D printing technology, you may need to generate volume supports or bar supports.

Controlling the size and shape (and therefore, the strength) of these supports plays a critical role in whether your prints will be successful or not. In this chapter, we explored how to generate volume supports and bar supports for FFF prints. We discussed how to effectively apply these supports to the entire body or selected surfaces within our components. We showcased how to apply supports to both solid and mesh bodies. We demonstrated the impact of changing support settings as well as slicing settings to control how those supports are sliced and toolpaths are generated. We touched on the impact of the gap between our parts and the support structures when generating supports so we can print a successful down-skin region while reducing the post-processing workload of our parts.

We also highlighted how to create bar supports for SLA/DLP printing. We showcased the similarities between bar supports for FFF and resin-based printers as well as the unique properties of bar supports for SLA /DLP printers, such as tree structures, grouping, and bracing. We also saw how to create base plate supports in order to achieve a strong adhesion to the build plate.

Using the techniques you have learned in this chapter, you can experiment with your specific printer, material, and print settings in order to create support structures that are just as strong as necessary to successfully print a given part, while minimizing material usage and the post-processing time required to remove the printed support structures.

In the next chapter, we will further explore how to slice our models and generate an additive toolpath, along with how to simulate that toolpath layer by layer in order to better understand the effects of our model orientation, our slicing settings, and our support structures.

11

Slicing Models and Simulating the Toolpath

Welcome to *Chapter 11*. In this chapter, we will learn all about how to generate and simulate the additive toolpath for **Fused Filament Fabrication** (**FFF**) 3D printing. We will go over how to acquire key information about our additive setup, such as print statistics and energy consumption. We will also learn how to generate and customize the G-code necessary to start a print job.

Once all the components we wish to print are properly oriented, arranged, and supported, it is time to generate the additive toolpath. After we generate the toolpath, we can visualize the outcome at each layer and simulate the movement of the nozzle during travel and extrusion moves.

Generating the additive toolpath is a one-click solution in Fusion 360, but it is always a good idea to review the toolpath to understand whether the print settings and body presets we have chosen will result in the desired outcome.

In this chapter, we will cover the following topics:

- Generating an additive toolpath
- Simulating an additive toolpath
- Generating G-code

By the end of this chapter, you will have a better understanding of how to generate an additive toolpath using Fusion 360. You will learn how to customize Fusion 360's preferences to generate a toolpath manually and automatically. You will learn how to access key print statistics, such as how much filament you will use to print a given print and how long the print will take. You will also learn how to calculate the total cost of a given print job, including both the filament cost and the cost of energy. You will learn how to simulate the additive toolpath to visualize the impact of your print settings on the print job layer by layer. Finally, you will learn how to generate a G-code file. As we get to the end of the chapter, you will acquire the necessary skills to incorporate a pause action at a specific layer during the print, in order to embed objects in your parts, such as magnets, hex nuts, and strengthening rods. You will also understand how to customize your G-code to change the filament used mid-print, in order to load a different color filament for a portion of your print.

Technical requirements

All topics covered in this chapter are accessible for the Personal, Trial, Commercial, Startup, and Educational Fusion 360 license types.

The lesson files for this chapter can be found here: `https://github.com/PacktPublishing/3D-Printing-with-Fusion-360`

In addition to Fusion 360, you will need a text editor such as Notepad to open the G-code file.

Generating an additive toolpath

In the previous chapters, we covered how to create an additive setup, orient and place our parts within the build volume, and generate support structures. In this section, we will build on this knowledge and demonstrate how to generate an additive toolpath for such an additive setup.

In order to demonstrate how to generate an additive toolpath within Fusion 360, let us use the `Connector-Assembly` provided within the GitHub page for this chapter. This Fusion 360 document consists of three components. As you can see in *Figure 11.1*, the main component is named `Connector`. There are two additional components named `Carbon Fiber Rod`. Based on how the parts are assembled, the carbon fiber rods are meant to be embedded within the Connector. In order to manufacture such an assembly using traditional manufacturing methods, we would have to build the Connecter in two halves, so we can embed the carbon fiber rods. Alternatively, we would have to drill access holes to place the carbon fiber rods within the connector body. However, as we will utilize additive manufacturing, we can simply insert the carbon fiber rods during the printing process without having to rely on secondary operations.

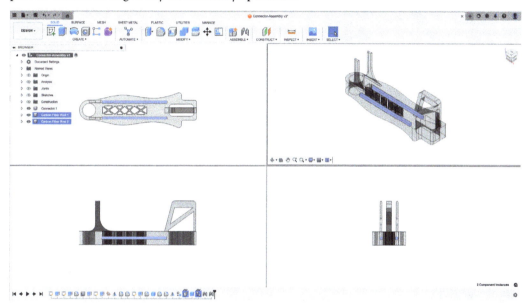

Figure 11.1 – Connector-Assembly with three components

Carbon fiber rods can easily be purchased from various sources and can be utilized in order to strengthen a given component in a certain direction. *Figure 11.2* shows that a quick search on www. amazon.com for carbon fiber rods will result in various sizes and prices available for purchase. In this example, we will rely on 4 mm by 4 mm by 420 mm rods, which come in packs of 8 for around $16.

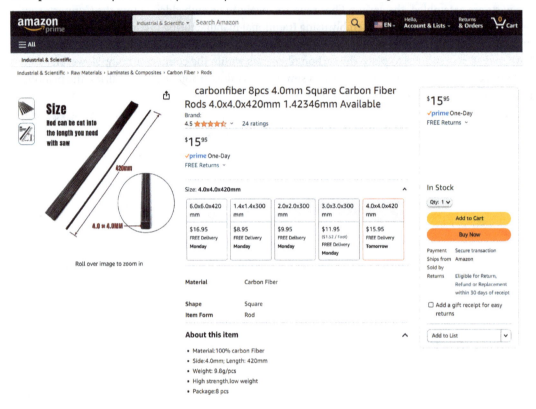

Figure 11.2 – Various sizes of carbon fiber rods are available online for purchase

The component named Connector already includes two cavities where the carbon fiber rods can be embedded during the printing process. The design can accommodate a 100 mm long rod with a 4 mm x 4 mm cross-section. As this example is simply a theoretical one, the cavity does not include additional tolerances. If we were printing this component as a part of our final design, it would be better to purchase the carbon fiber rods first and measure them, ensuring that their cross-section dimension is indeed 4 x 4 mm. Then, we would add one layer's worth of tolerance to their measured cross-sectional dimension in both the *Y* and *Z* axes, allowing us to easily place our rods within the part during the printing.

The next step in the process is to create an additive setup. We covered extensively how to create additive setups in the previous chapter. As you can see in *Figure 11.3*, the additive setup for this part includes a **Prusa i3MK3S+** for the machine, and a customized **PLA (direct drive)** for the print setting. If you look at the browser, under the active setup (Setup 1), you will see that the PLA print setting has a **Custom** body preset associated with it.

In the **SETUP** dialog, we will explicitly select our model as **Connector:1**. We will not choose carbon fiber rods as a part of our model selection process. This will ensure that we will only 3D-print the Connector part.

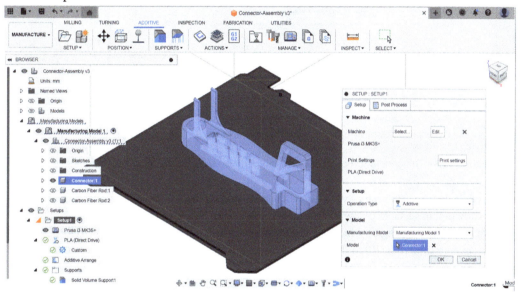

Figure 11.3 – Creating an additive setup for the component named Connector

At this point, I would like to highlight a key Fusion 360 preference we all should be aware of. You may recall that we can access the **Preferences** dialog by clicking on the Avatar icon in the top-right corner of the user interface, as shown with callout **1** in *Figure 11.4*. Within the **Preferences** dialog, you can navigate to the **Manufacture** section in the browser, as highlighted by callout **2**. In the **Manufacture** section, you will notice that there is a **Generate Additive Toolpath on Setup creation** option, as highlighted by callout **3**. By default, this option is unchecked, as shown in the figure.

If you often work with models that do not require support structures, you may prefer to activate this checkbox, and Fusion 360 will automatically create the additive toolpath for you as soon as you create an additive setup. I often find myself working with models that require the generation of support structures. Therefore, my preference is to leave this option unchecked. This way, I can first add the support structures and then generate the additive toolpath once I am satisfied with the support structures for my models.

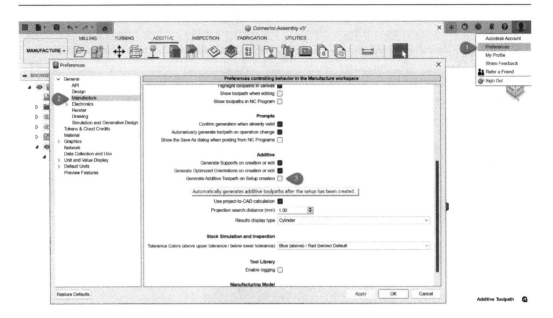

Figure 11.4 – The generation of an additive toolpath can be automated within Preferences

Once an additive setup has been created, we will need to add the necessary support structures for this model to down-skin surfaces below a certain overhang angle. *Figure 11.5* shows the selection of two faces with large overhang angles, which will require support structures in order to successfully 3D-print them. As covered in *Chapter 10* in detail, we will apply solid volume supports to these two faces.

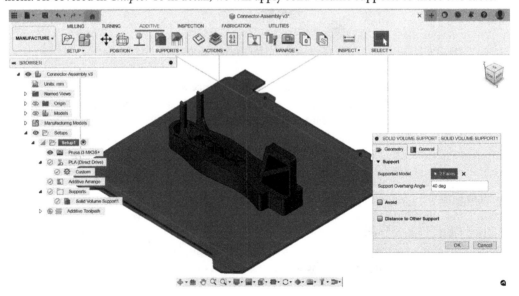

Figure 11.5 – Applying solid volume supports to select faces

Note that we did not apply support structures to the two cavities where the carbon fiber rods will be placed within our part. This is a key benefit of being able to select CAD-based surfaces when generating support structures. Using an explicit selection set, you not only have full control over which surfaces you wish to support; you can also easily avoid creating supports on portions of your model where you know you will perform mid-print operations, such as inserting magnets, hex nuts, or strengthening rods, which will ensure the part is self-supported.

Now that our setup is complete, it is time to create the additive toolpath. Fusion 360 allows you to create an additive toolpath in one of two ways. After selecting the additive toolpath within the browser, you can right-click and select **Generate** from the pop-up menu, as shown in *Figure 11.6*. Alternatively, after selecting the additive toolpath from the browser, you can select the **Generate** command within the **ACTIONS** panel located in the **ADDITIVE** tab. Both of these commands will start to generate the additive toolpath.

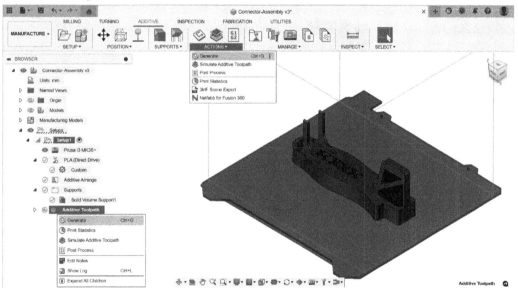

Figure 11.6 – Generating the additive toolpath

Once you have generated the additive toolpath for an FFF printer, if you expand the **Additive Toolpath** line item within the browser, you will see a line item called **Print Time**. This entry, which is highlighted with callout **1** in *Figure 11.7*, displays the total print time for your additive setup. You can acquire additional information by simply double-clicking **Print Time** within the browser, or by selecting the **Print Statistics** command under the **ACTIONS** panel, as indicated by callout **2** in the same figure.

Both of these actions will display the **PRINT STATISTICS** dialog, which consists of two tabs. The first tab, named **Print Statistics**, gives you additional information, such as how many meters of filament will be used to successfully finish printing this setup. This is an important piece of information, as you may need to make sure you have enough material at hand before starting your print. You'll also see information about how many layers the print will consist of, and how much time will be consumed by various portions of your model print, such as its shell, infill, and support structures.

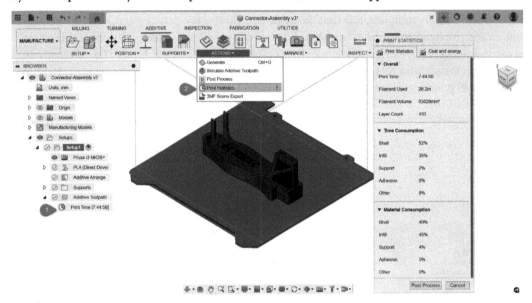

Figure 11.7 – Print statistics for an additive setup

In the second tab of the **PRINT STATISTICS** dialog, you can calculate the material cost and the amount of energy you will consume in order to finish printing this part. Note that the **Cost and energy** tab is only available for FFF machines. You will not be able to inquire about the material or energy costs for other additive manufacturing processes.

As shown in *Figure 11.8*, the **Cost and energy** tab allows you to select your location from the drop-down menu. Based on your location, you can enter the cost of energy in your market and the cost of the specific filament you have purchased for this print.

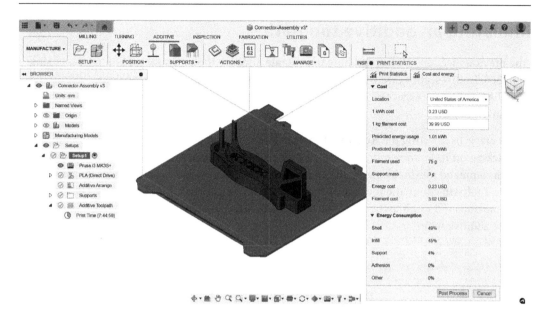

Figure 11.8 – The cost and energy calculation for an additive setup with an FFF machine

In this example, we can set the location to the USA and enter the cost of energy as $0.23 per kWh, which is the current average cost of energy in the US at the time of writing. We can enter the cost of filament used as $39.99 per kilogram. With these inputs, Fusion 360 estimates that the total cost of the print will be $3.25, based on how much filament we will use and how much heating will be involved in extruding the filament, as well as keeping the bed at a certain temperature. Note that this calculation does not include the cost of the carbon fiber rods we will insert into this part.

If you recall, the cost of 8 rods was $16 on www.amazon.com. This means each rod is worth $2. Each rod was 420 mm long, and our openings were 100 mm long. This means we can use two-quarters of a rod per part. So, we will use $1 worth of rod per part, bringing the total cost of this part to $4.25, excluding the labor cost.

Being able to see such information prior to printing helps us understand the approximate cost to manufacture our parts. Based on this information, we can choose to make a change to our print settings to reduce either the energy or the filament cost for the final part.

In this section, we highlighted how to generate an additive toolpath automatically and manually. We pointed out the preferences you can customize in order to generate the additive toolpath as soon as you create an additive setup. We showcased how to generate the additive toolpath from the browser and the ribbon UI of Fusion 360. We also learned how to use Fusion 360 to calculate critical print statistics, such as the total print time, material required, and print cost. In the following section, we will simulate the additive toolpath to visualize it layer by layer.

Simulating an additive toolpath

In the previous section, we learned how to generate additive toolpaths for an FFF process using Fusion 360. In this section, we will focus on how to simulate these additive toolpaths. Within the **MANUFACTURE** workspace of Fusion 360, we have two paths we can follow in order to simulate the additive toolpath.

The first one is by selecting the **Additive Toolpath** line item in the browser for the active setup and right-clicking on it. As shown in *Figure 11.9* with callout **1**, we can access the **Simulate Additive Toolpath** command within this right-click menu. The second option we have is to utilize the **Simulate Additive Toolpath** command located within the **ACTIONS** panel of the **ADDITIVE** tab. Both of these commands will result in the same outcome, which is the simulation of the additive toolpath for the active additive setup.

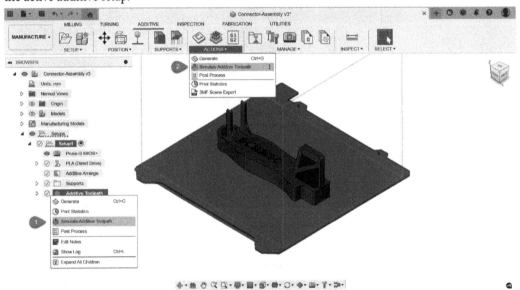

Figure 11.9 – How to access the Simulate Additive Toolpath command

Once we simulate the additive toolpath, we can visualize all the layers our machine will 3D-print for this additive setup. There are four ways we can change the layer number we visualize. *Figure 11.10* points out each of these methods with a unique callout:

- Callout **1** points to a blue arrow within the canvas of Fusion 360. If we select this arrow by left-clicking on the mouse, we can drag the arrow up or down along the Z axis to visualize the toolpath at a different layer.

- Callout **2** points to the simulation player controls located at the bottom of the canvas. There are seven buttons in this control. Using the play button in the center of the control (button number four from the left), we can start the animation of the toolpath. The leftmost and rightmost buttons (buttons one and seven from the left) within the control will automatically jump to the first and the final layer. In this example, the final layer is **433**. We can also utilize buttons two or six in order to toggle from one layer to another at a time. Buttons three and four animate the toolpath within the active layer.

- Callout **3** points to the **Current Layer** input field within the **SIMULATE ADDITIVE TOOLPATH** dialog. *Figure 11.10* shows the toolpath for the first layer, as the input for the Current layer is set to 1. We can type a specific layer number into this field to quickly jump to that layer's toolpath visualization. Keep in mind that you do not need to press *Enter* or click **OK** for the change to take effect after typing in the desired layer number. Fusion 360 will automatically update and display the toolpath for the layer you have entered.

- Callout **4** points to the Simulation player timeline, which is located below the Simulation player controls. The black vertical line can be used to control the timeline. By selecting it with a left-mouse click, you can drag the timeline toward the right in order to visualize the toolpath for any of the calculated layers.

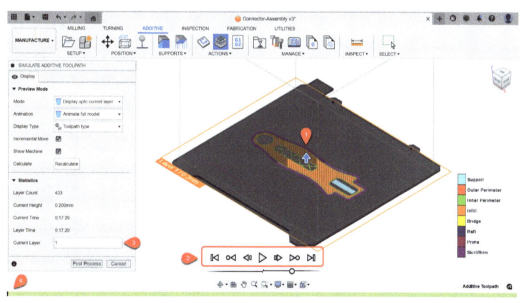

Figure 11.10 – How to change the layer being displayed within the toolpath simulation

The **SIMULATE ADDITIVE TOOLPATH** dialog has several options that allow you to visualize the toolpath specific to your needs. If you change the **Mode** input from **Display upto current layer** to **Display current layer only**, you will be able to visualize the toolpath specific to the layer number you are interested in. In *Figure 11.11*, we can see the toolpath for our Connector component at layer number 80. You can also see that the **Display Type** option is set to **Toolpath Type** within the **SIMULATE ADDITIVE TOOLPATH** dialog. With this display type, Fusion 360 colors each toolpath line according to the type of tool path it is creating. For example, any outer perimeter toolpath is colored with a red line, whereas all inner perimeter toolpaths are colored in green. Support structures are represented using a light blue line, and the infill toolpath is shown as orange lines. Those are the only four types of toolpath present in the layer shown in this figure.

You may have also noticed that the **Show Machine** checkbox is unchecked in *Figure 11.11*, and we are looking at our model from the top view. Turning off the machine visualization hides the build volume as well the build plate for the 3D printer we are utilizing in the active setup. By hiding the visualization for the build plate, we can easily focus on our parts and supports while simulating the additive toolpath.

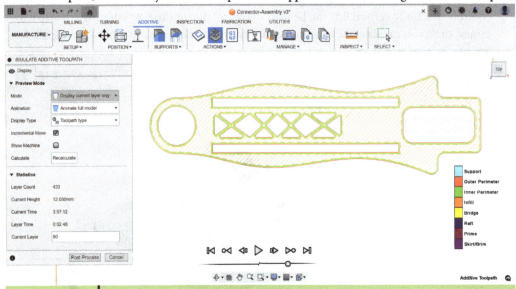

Figure 11.11 – Visualizing the toolpath for a specific layer

Another option we have in the **SIMULATE ADDITIVE TOOLPATH** dialog is to display the print speed for our toolpath. We can do that by changing the display type from **Toolpath type** to **Print speed**, as shown in *Figure 11.12*. When visualizing our toolpath using the print speed display type, Fusion 360 colors each toolpath based on the extrusion speed utilized while 3D-printing each toolpath. The image on the left shows that our part and support structures are printed using a relatively slow speed for layer 1. The image on the right shows the same parts at layer 81. Here, we can see that our infill, perimeter, support, and bridge toolpath types are printed at different speeds, and they are generally much faster on this layer compared to layer 1.

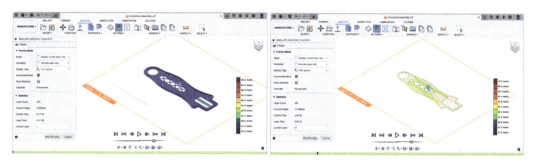

Figure 11.12 – Displaying an additive toolpath using a print speed display type

By simulating an additive toolpath and visualizing what our 3D printer will do at each layer while extruding a filament, we can be sure of how many perimeter toolpaths our part will have at various layers and how our parts will be supported during the printing process. We can also visualize the change in print speed from one layer to another.

Using these techniques, we can either choose to proceed with the print or change certain print settings in order to achieve our desired toolpath for perimeters, infills and supports. In the next section, we will learn about how to generate G-code in order to start a print.

Generating G-code

After creating our additive toolpath and simulating it, if we are satisfied with the outcome, we can move on to the final step in the process, which is the generation of the G-code. In order to generate G-code using Fusion 360 for an additive setup, we can simply click the **Post Process** button at the bottom of the **SIMULATE ADDITIVE TOOLPATH** dialog, as shown in *Figure 11.13*. This action will automatically bring up the **NC Program** dialog, where we can post-process our toolpaths and create the necessary G-code.

Within the **NC Program** dialog, there are two tabs. The first tab, called **Settings**, is the only relevant tab for additive manufacturing. So, we can simply ignore the **Operations** tab. The **Settings** tab of the **NC Program** dialog has two columns. The column on the left allows us to choose which post-processor we wish to use in order to convert the toolpath we see on our screen to a G-code file. The column on the right, named **Post Properties**, allows us to customize the output so that the G-code we produce behaves specific to our instructions.

Within the **Program** subsection of the **NC Program** dialog, we can name our G-code file and add additional comments. We can also designate the output folder for the G-code file. If we select the **Post to Fusion Team** checkbox, as shown in *Figure 11.13*, our G-code file will not only be saved in our local folder but we'll also create a copy of it within the **FUSION TEAM** folder of our choosing.

Figure 11.13 – Post-processing an FFF additive toolpath

Another way we can initiate the **NC Program** dialog is by selecting the **Post Process** command within the **ACTIONS** panel of the **ADDITIVE** tab, as shown in *Figure 11.14*. Please remember that you have to select the additive toolpath for the active setup prior to selecting the **Post Process** command in order to initiate the **NC Program** dialog.

In this example, we worked with a Prusa i3MK3S+ machine. The post file utilized for Prusa printers has additional properties, as shown in *Figure 11.14*. The property we are interested in exploring in this example is in the **Pause Print** section, as illustrated with callout **2**. By expanding the list of the **Pause Print** section, we can enter a layer number at which we wish to pause the printing process. We can also include a message to display on the printer's display screen during the pause.

As you may recall from the previous section, by simulating our additive toolpath, we identified that at layer 80, the extruder will attempt to bridge over the cavity where we wish to add our carbon fiber rods. Therefore, it is a good idea to pause the print at layer 79 so that we can insert our carbon fiber rods before the printer attempts to encapsulate the cavity.

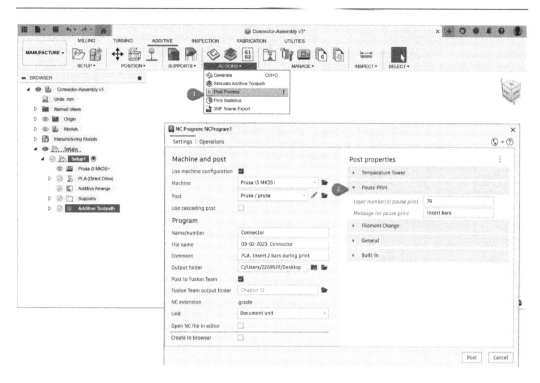

Figure 11.14 – Adding a Pause Print function to the G-code at a specific layer

Fusion 360 offers unique post properties based on the Post file selected, which is associated with the machine you have selected for your setup. *Figure 11.15* shows the differences between three common 3D printers and their post properties.

The image on the left shows the post properties for the Prusa family of printers. As you can see in the figure, using a Prusa post, you can customize your G-code so that you can find the optimum temperature to print your specific filament. This is possible by enabling the **Trigger** field within the **Temperature Tower** section. This action will enable you to print a test object while controlling the rise in temperature, from an initial temperature value to a final one with a set of intervals. After the print, you can observe the finished product and decide at which temperature you get the best outcome, using that knowledge in subsequent prints as a new print setting. As we highlighted previously, in *Figure 11.14*, you can also execute a **Pause Print** function. In addition to these two sections, you can also execute a **Filament Change** operation by entering the layer number at which the filament should be swapped for a different one.

The middle image within *Figure 11.15* is from an Ultimaker post file for an Ultimaker S5 printer. In the **General** section of **Post properties** for this post file, you also have the ability to indicate which material and print core is loaded onto the machine for Extruder 1 and Extruder 2. Without this information, you would not be able to print successfully on this printer.

The image on the right within *Figure 11.15* is from a Creality post file for a Creality CP-01 printer. Just like the Prusa and Ultimaker families of printers, Creality printers have the **Temperature Tower** options under **Post properties**. In addition, you can control whether to enable or disable bed leveling. You also have additional functionality to control at which *Z* height to turn off the bed leveling compensation.

Figure 11.15 – Post properties based on manufacturer

Another very useful functionality within the **NC Program** dialog is the ability to edit an expression for a given field. *Figure 11.16* shows how you can customize any input field, such as **File name**, beyond simply entering the input manually. If you select the : icon to the right of the input field, as shown with callout **1**, you can select **Edit expression** from the drop-down menu, as shown with callout **2**. When the **Expression** dialog is displayed, you can enter parameters such as `date` or `documentName` so that the file name used for the G-code is automatically set to the parameter you have designated. In the **Expression** dialog, you can also use some basic formulas.

Figure 11.16 demonstrates how to concatenate multiple parameters by using the + sign. You can also add custom strings between the `date` and `documentName` parameters by using _ between the two. After selecting **OK**, the **File name** field will be automatically populated with the date and the name of the Fusion 360 document. The **Expression** dialog also has the ability to auto-fill parameter names. If you are not familiar with the all available parameters, you can simply start typing an input, such as name, and you will see various parameters containing the string `name` so that you can choose a parameter, such as the Fusion Team name or document name, from a list of available options. Once your expression has been set, you can store it by selecting **Save as the user default** so that the next time you create an NC program to export G-code, you don't have to type in the name of the file, and your filename will be automatically set as a combination of the date and the Fusion 360 document name.

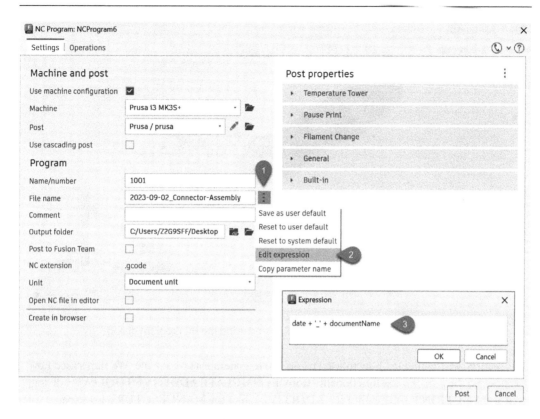

Figure 11.16 – How to utilize expressions for input fields in NC Program

Once you are done customizing your post properties, all you have to do is select the **Post** button in the bottom-right corner of the **NC Program** dialog. Fusion 360 will then export a G-code file in the location you have designated with the filename you have entered. *Figure 11.17* shows the output of this action, which is a human-readable text file with a `.gcode` extension. If you open this file using your native text editor app, such as Notepad, and scroll down, you will be able to spot the beginning of layer 79. As you can see in *Figure 11.17*, a pause print action is introduced with the code `M601`, as well as a display message with `M117`, with text that reads `Insert Bars`. If you print this G-code on this 3D printer, the print will proceed until it reaches layer 79, and it will pause until you press the knob. Then, you insert your carbon fiber bars.

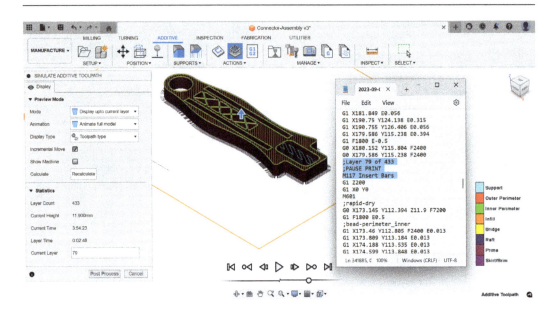

Figure 11.17 – Visualizing the G-code posted using the NC Program dialog

In this section, we highlighted how to use Fusion 360 to generate a G-code file. We showcased how you can access the post-process functionality from the **SIMULATE ADDITIVE TOOLPATH** dialog, as well as the **ACTIONS** panel within the **ADDITIVE** tab of the **MANUFACTURE** workspace. We covered the various options available in the **NC Program** dialog to convert our additive toolpath to a G-code file so that we can print it. We also highlighted the various post properties available within the **NC Program** dialog based on our selected 3D printer.

Summary

In this chapter, we learned all about how to slice our models for an FFF 3D printer. Slicing is an important step in print preparation for additive manufacturing and comes after creating a setup, orienting parts, arranging them on the build platform, and generating support structures. Once our parts are sliced, we can also simulate the additive toolpath.

In this chapter, we learned how to generate and visualize additive toolpaths layer by layer. We went over the various print statistics, such as energy consumption, print time, material usage Fusion 360 displays for FFF printers. With all this information at our fingertips, we have the power to post-process the additive toolpath as is and create a G-code file. We also highlighted how we can change our print settings based on the simulation results and print statistics, enabling us to print our parts with less material and use less energy.

In the final section of this chapter, we learned how to post-process the additive toolpath and create G-code. At the time of G-code creation, we also learned how to customize it. Using these customizations, we learned how to create temperature towers so that we could determine what temperature to print using certain filaments. We also learned how to pause a print in order to insert additional objects, such as hex nuts, magnets, and strengthening rods. We also learned how to change the filament mid-print so that we could print different segments of our parts using different filament colors. Finally, we learned how to automate certain inputs within the **NC Program** dialog, such as the filename, using parameters that are already available within a Fusion document. Using all these techniques, we can confidently go from an additive setup to generating G-code for an FFF printer.

A lot of the fundamentals of positioning and supporting components for metal 3D printers are similar to what we learned for FFF 3D printing. However, there are some process-specific settings we will need to go over for metal printers. In the following chapter, we will switch gears and cover more about 3D printing with metals.

Part 4:
Metal Printing, Process Simulation, and Automation

This final part of the book is all about metal 3D printing and automation. First, we will introduce the various ways you can use Autodesk Fusion to create an additive setup for a metal 3D print. We will explain how you can utilize fused filament fabrication, binder jetting, directed energy deposition, and metal powder bed fusion technologies to create metal parts. Then, we will demonstrate how to simulate the metal powder bed fusion process in order to detect potential print failures. We will also demonstrate how to compensate for large distortions in our models and ultimately minimize possible print failures.

We will end the book by highlighting various automation techniques that we can utilize to reduce the time spent on repetitive tasks. We will introduce premade automation tools available for download and use on the Autodesk App Store. We will introduce the concept of operation templates as well as scripts created with Python or C++, using Autodesk Fusion's API, and how to customize them to fit our needs.

This part has the following chapters:

- *Chapter 12, 3D Printing with Metal Printers*
- *Chapter 13, Simulating the MPBF Process*
- *Chapter 14, Automating Repetitive Tasks*

12

3D Printing with Metal Printers

Welcome to *Chapter 12*, where we will learn all about metal additive manufacturing. Metal 3D printing has been increasing in popularity over the past several years as it offers a long list of benefits for numerous industries, such as aerospace, healthcare, and high-end consumer goods. Using metal 3D printing, you can create parts with complex internal structures and organic external faces, which are impossible to manufacture any other way. 3D printing with metal is also more cost-effective compared to other manufacturing methods for low quantities.

Fusion 360, along with the Manufacturing extension, can be utilized to create additive setups with 3D printers that are capable of manufacturing parts using metals. In this chapter, we will introduce the various ways you can use metal additive manufacturing within Fusion 360. We will go over how to utilize Fusion 360 to prepare a model to be 3D printed using fused filament fabrication, metal binder jetting, metal powder bed fusion, and directed energy deposition techniques. We will also show how to select associated print settings based on the printing technology. We will highlight how to orient and place our parts based on machine-specific factors, such as recoater blade direction.

Depending on the additive technology utilized by our specific 3D printer, as well as our part orientation, parts may need support structures during the printing process. As the supports utilized during metal powder bed fusion are different compared to the supports required for fused filament fabrication and resin printing, we will also learn about how to create technology-dependent support structures to minimize material waste, decrease print time, and minimize post-processing effort.

Whether we are utilizing a customized **Fused Filament Fabrication** (**FFF**) 3D printer to print with a metal filament, or we use a metal binder jet 3D printer, we may need support structures to successfully print our parts or to provide the necessary support for our parts during the sintering process. In this chapter, we will also learn how to use Fusion 360 to create setter supports to prevent our parts from deforming due to the high temperatures they will be exposed to within a sintering oven.

In this chapter, we will go over the following topics:

- Introduction to metal 3D printing

- 3D printing with metal powder bed fusion machines

- Setter supports for a successful sintering process

By the end of this chapter, you will have a good understanding of the various 3D printing technologies you can use to manufacture parts using metal materials. You will have learned how to create an additive setup using Fusion 360 with a 3D printer capable of manufacturing a metal part. You will gain an understanding of how to choose an appropriate part placement and orientation based on your 3D printing technology. You will have also learned how to create lightweight support structures for metal 3D printing. Finally, you will have learned how to create support structures for parts printed using metal binder jetting to minimize distortions during the sintering process.

Technical requirements

In this chapter, we will cover metal additive manufacturing. We will highlight how you can use FFF and binder jetting printers to 3D print in metal. These printers are accessible to all Fusion 360 users (license type: Personal, Trial, Commercial, Startup, and Educational).

However, we will also utilize multiple **metal powder bed fusion** (**MPBF**) 3D printers and their respective print settings. We will highlight how to generate support structures for MPBF machines and metal binder jet machines. These support structures are not available with Fusion 360 until you gain access to the Manufacturing extension.

As with all other Fusion 360 extensions, the Manufacturing extension is not available for personal use.

If you have a commercial license of Fusion 360, to create an Additive setup with an MPBF machine, you will also need access to the Manufacturing extension (`https://www.autodesk.com/products/Fusion-360/manufacturing-extension`)

Autodesk offers a 14-day trial for the Manufacturing extension for all commercial users of Fusion 360. You can activate your trial within the extensions dialog of Fusion 360.

The Manufacturing extension is also available for free to Startup and Educational license users.

The lesson files for this chapter can be found here: `https://github.com/PacktPublishing/3D-Printing-with-Fusion-360`.

Introduction to Metal 3D printing

In *Chapter 7*, we created our first additive setup using an FFF printer. In this chapter, we will highlight how to create additive setups with machines that are capable of 3D printing with metal.

The Fusion 360 machine library includes machines with additive capabilities for various technologies. *Figure 12.1* shows how we can filter additive machines based on their technology within the machine library. Fusion 360's machine library can be accessed within the **MANUFACTURE** workspace. You can access the **Machine Library** command within the **MANAGE** panel of any tab. If you have access to the Manufacturing extension, you will be able to select an MPBF machine within **Machine Library** while creating an Additive setup:

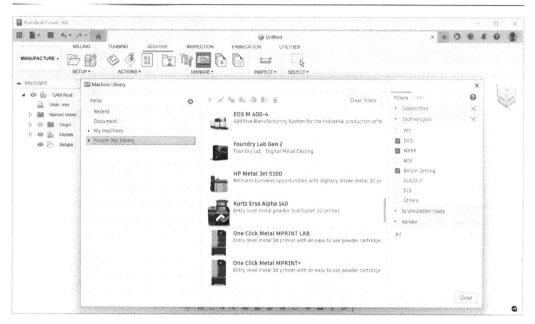

Figure 12.1 – Fusion 360 Machine Library

As we mentioned in the introduction to this chapter, multiple technologies are capable of generating metal parts when using additive manufacturing. You can create a metal part using FFF, MPBF, binder jetting, **directed energy deposition (DED)**, and **electron beam melting (EBM)**. Fusion 360 can help you generate metal parts using four of the five additive manufacturing technologies, as shown in *Table 12.1*. For each of the four supported additive manufacturing technologies, Fusion 360 has machines from a list of machine manufacturers that you can use to create your Additive setup. Each machine vendor and technology has its capabilities and limitations, so the workflow to create metal parts will differ from technology to technology and vendor to vendor.

In this chapter, we will mainly focus on MPBF and binder jetting technologies. However, in this section, we will briefly introduce how you can use FFF to create metal parts using Fusion 360:

Fused Filament Fabrication	Directed Energy Deposition	Binder Jetting	Metal Powder Bed Fusion
Ultimaker *	Autodesk Generic	Desktop Metal *	Aconity
Raise3D *	Meltio	Sintertek *	Additive Industries
All FFF machines (0.6 mm nozzle)		HP	DMG Mori
		Digital Metal	EOS
			Kurtz Ersa
			One Click Metal

Fused Filament Fabrication	Directed Energy Deposition	Binder Jetting	Metal Powder Bed Fusion
			Renishaw
			SLM Solutions
			Xact Metal

Table 12.1 – Fusion 360 supported machine manufacturers for metal 3D printing

In *Table 12.1*, you will notice that certain manufacturers have an * next to them. The Fusion 360 machine library includes machines from all of the machine manufacturers listed in this table. However, you may not be able to use Fusion 360 to slice your models and create the additive toolpath to 3D print them using a metal filament or metal powder for all machines in **Machine Library**.

For machines from manufacturers with an *, Fusion 360 may not have the necessary print setting in its library or may not be able to create the specific slicing file format required by the machine maker. You can, however, use Fusion 360 to orient your parts within the printer's build volume and generate support structures as needed.

Both UltiMaker and Raise3D utilize FFF technology to 3D print with metal filaments. If you have an UltiMaker S5 (or Method 3D) printer with the metal expansion kit, you can upgrade your 3D printer to unlock metal FFF capabilities. Using this upgrade, you will be able to print using BASF Ultrafuse 17-4PH or 316L material. The necessary print settings for these materials can be downloaded from UltiMaker Cura software's Material Marketplace. Once downloaded, you can use them with cc0.4 or cc0.6 cores, as shown in *Figure 12.2*. The Fusion 360 print setting library does not include a validated print setting for these metal filaments with UltiMaker printers:

Figure 12.2 – UltiMaker S5 with metal expansion kit and filaments from BASF

If you have access to a Forge 1 printer from Raise3D, you can use the same BASF filaments (17-4PH of 316 L) to generate metal parts using your FFF machine. Similar to 3D printing metal parts with UltiMaker S5 machines, the Fusion 360 print setting library does not include a validated print setting for 17-4PH or 316L from BASF for Raise3D printers.

In the case of Forge 1, you will need to use Raise3D's native slicer `ideaMaker for Metal` to slice your models and generate your additive toolpaths as the idea Maker software contains the necessary print profiles for these materials.

Only UltiMaker and Raise3D offer a dedicated solution to 3D print with stainless steel filaments, debinding, and sintering solutions. In the case of UltiMaker, you need to ship your printed metal parts to one of several dedicated debinding and sintering service providers to get your finished part in the mail. If you have a Raise3D Forge 1 printer, you can also purchase a debinding and sintering station from Raise3D to complete the entire process at your facility.

If you have an FFF printer from a different machine provider, you can use metal filaments, specialized extruders, as well as debinding and sintering stations from solution providers such as The Virtual Foundry (`https://shop.thevirtualfoundry.com/`). As you can see on The Virtual Foundry's web page, they currently offer eight metal filaments – Aluminium 6061, Bronze, Copper, H13 Tool Steel, High Carbon iron, Inconel® 718, Tungsten, Stainless steel 17-4 and 316L, as well as Titanium 64-5 – to their customers.

The Fusion 360 print setting library already contains the necessary print settings for Copper and Bronze from The Virtual Foundry. The following blog from Autodesk highlights how designers are already using this method of metal 3D printing with Fusion 360 and Prusa 3D printers: `https://www.autodesk.com/products/fusion-360/blog/imagining-3d-printed-futures-ts-technical-design-virtual-foundry/`.

Please keep in mind that if you choose to use this workflow, you will have to manually scale up your parts by a predesignated multiplication factor for each material before slicing and generating the additive toolpath. Scaling before 3D printing is required because parts shrink during the sintering process. In the previously linked Autodesk blog, the end user scaled up their parts by 10% before slicing and printing them.

An example of this workflow can be seen in *Figure 12.3*. The Fusion 360 document named `Slotted Liner` is also provided in the GitHub repository for this chapter. This Fusion document shows how you can set up a metal additive print using a Prusa i3 MK 3S + printer. The following figure shows an Additive setup with a customized Prusa printer. As you can see in the browser, this Additive setup uses a 0.6 mm nozzle instead of the default 0.4 mm nozzle. This customization has been done by editing the 3D printer selected for this Additive setup. Additionally, this setup includes a **Virtual Foundry Copper (Direct Drive)** print setting. In the same figure, you can see **Print Setting Editor** for this material, which displays the **Infill** subception for the **Normal** body preset. As we can see, this print setting uses a `100%` infill density with a `Rectilinear` pattern. The Additive setup is based on the component within **Manufacturing Model 1**.

If you open this model in Fusion 360 and look at **Manufacturing Model 1**, you will also notice that the component being 3D printed was scaled up 10% versus the model in the **DESIGN** workspace. You may recall that we covered how to modify components within the **MANUFACTURE** workspace using Manufacturing Models back in *Chapter 6*:

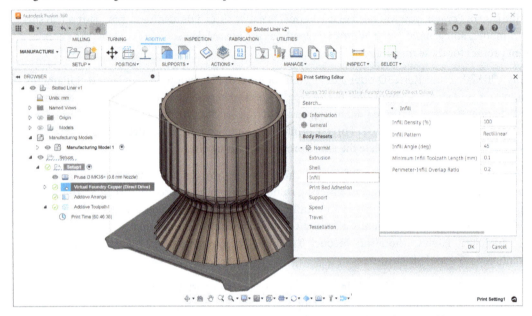

Figure 12.3 – Prusa i3MK3S+ with a 0.6 mm hardened steel nozzle and copper filament

Now that we have covered the 3D printing options with metal filaments using FFF printers, let's move on to another technology: DED. In addition to a generic DED printer, the Fusion 360 library of machines includes a DED printer from the machine maker MELTIO named M450. This printer utilizes multiple lasers and metal wire to manufacture parts. You can create an Additive setup with this printer and choose the 316SS print setting from Fusion's print setting library to generate the additive toolpath necessary for this printer.

As the DED process does not require debinding or sintering, there is no need to scale your components up before creating your Additive setup. After generating an additive toolpath and simulating it, as shown in *Figure 12.4*, you can post-process the additive toolpath and generate the necessary G-code to manufacture your parts using this 3D printer. *Figure 12.14* showcases an Additive setup of six metal watch casings in Fusion 360 and a corresponding 3D Print using this printer.

The image source for the printed parts can be found at `https://meltio3d.com/technology/`:

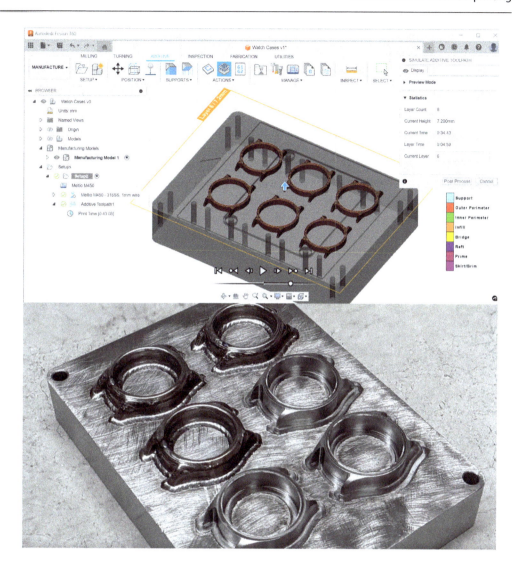

Figure 12.4 – Additive setup with a Meltio M-450 and the printed parts

Another technology you can utilize with Fusion 360 to 3D print parts using metals is binder jetting. You can create Additive setups for several Desktop Metal printers, as well as the Sinterjet M60 printer from Sintertek using Fusion 360. Within these setups, you can orient and arrange your parts using automatic orientation tools and the Additive Arrange functionality we have covered in previous chapters.

Once your additive setup is complete, you will need to export your setup as a 3MF file and use the native slicers from these companies to 3D print your parts. Depending on which machine manufacturer you are working with, you may also need to scale your parts up before generating the 3MF file based on the material you are interested in 3D printing.

If you are printing with a metal binder jet printer from HP, Fusion 360 offers additional functionality for connecting Fusion to your printer and sending print jobs directly to your machine. *Figure 12.5* shows an additive setup of a Connector part oriented and arranged for an HP Metal Jet printer.

In this Fusion 360 document, the **DESIGN** workspace contains a single component. **Manufacturing Model 1** was used to create the 70 components to be printed. Setup 1 shows the outcome of using the Additive Arrange functionality to position all 70 components within the build volume of this 3D printer. After installing and running the HP 3D Printers for Fusion 360 App in the Autodesk App Store (located at https://apps.autodesk.com/FUSION/en/Detail/Index?id=5739757793764591673), you will notice a new panel named **HP** appear within the **ADDITIVE** tab. Using the commands within this panel you can connect Fusion 360 to your HP Jet Fusion printers by entering your printer's IP address. You can also store and manage a list of all your HP printers within Fusion. After connecting to a specific printer, you can inquire about additional details from that printer and get the status of your previously submitted print jobs. Once you are ready to 3D print your models, you can also send your print jobs directly to the printer using this utility:

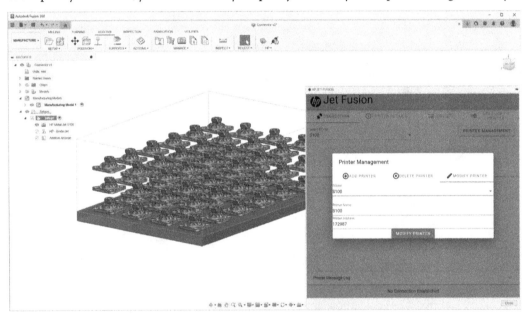

Figure 12.5 – Additive setup with an HP Metal jet S100 printer

In this workflow, you do not need to worry about scaling your models as the connection between Fusion 360 and HP Jet Fusion printers automatically scales your models based on your material selection, when you are sending your jobs to the printer.

The final binder jetting printer within Fusion 360's **Machine Library** we will highlight in this section is Digital Metal PX100. Within the **MANUFACTURE** workspace of Fusion, you can create a machine build file for this printer to start a print job. *Figure 12.6* shows the same model that we used in the previous figure within a new manufacturing model.

Manufacturing Model 2 has a total of nine copies of the same component. **Setup2** contains all the components within **Manufacturing Model 2** with a Digital Metal PX100 printer and a 316L Stainless Steel print setting. When you are creating an additive setup for this printer, you have to edit the manufacturing model and scale your parts up depending on the material you wish to 3D print with.

In this example, we scaled these nine components by a non-uniform scale factor within the manufacturing model to account for the uneven shrinkage during the sintering process. Once the additive setup is complete, to initiate the print, you can access the **Create Machine Build File** command located within the **ACTIONS** panel of the **ADDITIVE** tab. This will display the **CREATE MACHINE BUILD FILE** dialogue and allow you to select **CLI** as the export file format. Once you've created your slice file (`*.cli`), you can take it to your printer and start the print job:

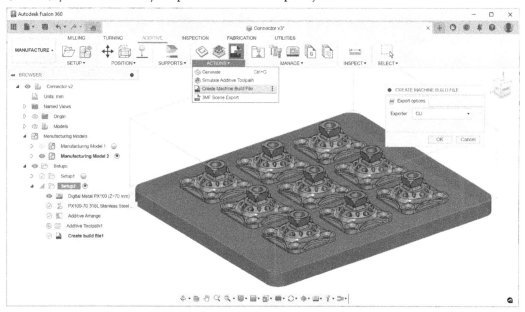

Figure 12.6 – Additive setup with a Digital Metal PX100 printer

In this section, we highlighted how to create setups for three of the four additive manufacturing technologies available within Fusion 360 to 3D print with metals. We covered how you can 3D print with metal with printers from UltiMaker and Raise3D. We also highlighted how you can customize your existing FFF printers and 3D print with filament from vendors such as The Virtual Foundry. In addition to FFF, we showcased how to utilize DED as well as Binder Jetting technologies with Fusion 360 to create additive setups for metal 3D printing. Both the FFF and Binder Jetting technologies require additional post-processing steps, such as sintering, to generate fully dense metal parts.

In the following section, we will introduce metal 3D printing using powder bed fusion technology. We will also talk about the unique support structures you can create using Fusion 360 for this additive manufacturing technology.

3D printing with metal powder bed fusion machines

MPBF is a fully supported additive manufacturing technology in Fusion 360 for 3D printing parts with metal powders. Fusion 360's **Machine Library** currently has MPBF printers from the following nine machine makers: Aconity, Additive Industries, DMG Mori, EOS, Kurtz Ersa, One Click Metal, Renishaw, SLM Solutions, and Xact Metal. Having access to the Manufacturing extension is a prerequisite to creating an additive setup with an MPBF 3D printer from any of these machine providers.

In addition to these nine machine makers, you can also use the `Autodesk Generic MPBF` machine to create an additive setup, as shown in *Figure 12.7*. After creating the setup and arranging your parts on the build platform, assuming you do not need support structures, you can generate the necessary machine files for 3D printers, which can utilize generic file formats such as SLC and CLI. You can access the **Create Machine Build File** command from the **ACTIONS** panel of the **ADDITIVE** tab within the **MANUFACTURE** workspace. The **CREATE MACHINE BUILD FILE** dialogue will allow you to specify the **Exporter** component you wish to use while creating the toolpath file for your printer:

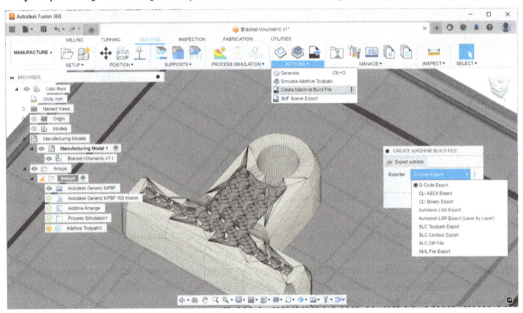

Figure 12.7 – Additive setup with an Autodesk Generic MPBF printer

Now that we have created an additive manufacturing setup with a generic MPBF printer, let's demonstrate how to create a similar setup with a 3D printer from a specific machine vendor.

In this next exercise, we will utilize a component designed with automated modeling in Fusion 360 to create an additive setup with a `One Click Metal MPRINT+` printer. After switching to the **MANUFACTURE** workspace, we can create an additive setup within the **ADDITIVE** tab. During the setup creation, when selecting our printer, we can select an associated print setting. In *Figure 12.8*, **Setup1** is created using a print setting named `OCM MPRINT+ SS316L 40 Micron ATU`. This is one of two print settings that are included with Fusion 360 for this printer. These two print settings are for demonstration purposes only. If you have a **One Click Metal (OCM)** printer, you will receive a print setting from your OCM sales team for the material you are 3D printing with. After selecting the printer and the print setting, you can select **OK** within the **SETUP** dialogue and create an additive setup:

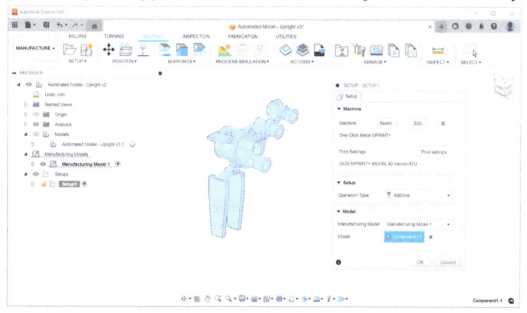

Figure 12.8 – Creating an additive setup with an OCM printer

Once the setup has been created, one of the first things we must do is arrange our parts within the build volume. For MPBF printers, we generally utilize the **2D Arrange** option, as shown in *Figure 12.9*. This arrangement type allows us to organize our parts within the build volume of our printer so that we can print multiple parts effectively. In this example, as we are only printing a single part, we don't have to worry about the object spacing input within the **ADDITIVE ARRANGE** dialogue. The 2D arrangement type will automatically place the selected component around the center of our build plate while making sure that our part is away from the edges of the platform at a distance, as indicated in the **Frame Width** input field.

We also want to make sure that our parts are above the build plate by a certain amount. In this example, we are using a 3 mm clearance between the build plate and our component. We need such a platform clearance so that our parts can be separated from the build plate using a bandsaw without damaging the printed model:

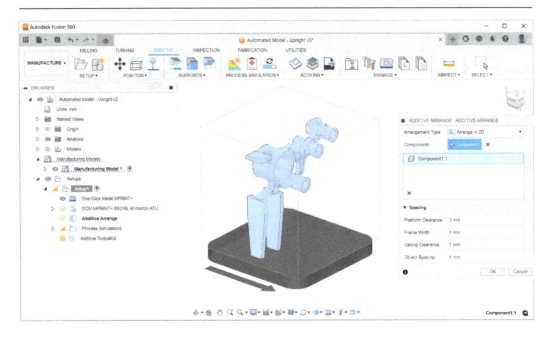

Figure 12.9 – Arranging a component within the build volume of a 3D printer

The next step in our print preparation for this MPBF additive setup is to orient our component. As you can see in *Figure 12.10*, we can use the **Automatic Orientation** command, which we covered extensively in *Chapter 8*, to find a suitable orientation for this part.

In this example, we also have to activate the **Support Bottom Surface** checkbox and enter a value of 3 mm for the **Distance to Platform (Z)** input field so that the outcome of our orientation study matches the desired arrangement outcome. The ranking inputs for an MPBF do not need to be modified if you are using the default ranking options Fusion 360 offers when utilizing the automatic orientation:

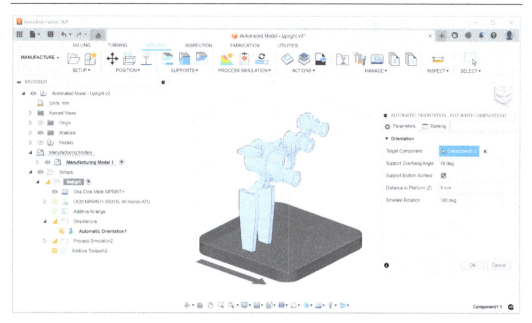

Figure 12.10 – Creating an automatic orientation study

Once the automatic orientation is complete, we can start looking through its results and toggle through different ranked outcomes. *Figure 12.11* shows the fifth-ranked outcome of this orientation result. In this example, we will utilize this outcome as it has the least amount of support volume necessary to print this part.

Another consideration when choosing an orientation result is the recoater blade direction. Fusion 360 displays the recoater blade direction with an arrow next to the build plate for most metal 3D printers, as shown in *Figure 12.11*. When printing parts with MPBF, thin horizontal segments of our parts tend to distort in the positive Z direction due to the thermally induced stresses on the layer being 3D printed. Such distortions create resistance between the recoater blade and the part while the 3D printer is applying a new coat of powder in between each layer. It is important to reduce such resistance between the part and the recoater blade so that we can print our parts successfully. If you are printing multiple parts within the same build volume, you need to consider their position on the build plate versus the recoater blade direction. You do not want your parts to be in a linear pattern perpendicular to the recoater blade direction. Instead, it is better to stagger them along the recoater blade direction to prevent a sudden resistance increase on the recoater blade and a crash between the blade and the parts:

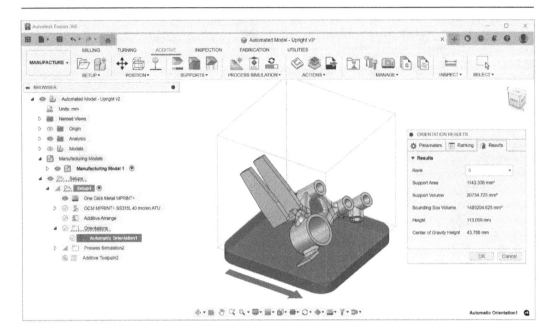

Figure 12.11 – Selecting an orientation to print the part

Up until this point, we have mostly repeated the print preparation steps we learned about in previous chapters. At this point, we will highlight some of the unique Fusion 360 capabilities we can utilize for MPBF printers after gaining access to the Manufacturing extension.

Figure 12.12 shows the list of support structures we can add to parts within an Additive setup with an MPBF printer. You may notice that this list is longer than the support structure we can add to FFF or SLA/DLP printers, which were covered in *Chapter 10*.

After gaining access to the Manufacturing extension, you will be able to add all the support structures Fusion 360 is capable of generating to any relevant additive setup. In *Chapter 10*, we mainly covered the **Bar** and **Base** plate supports. In this section, we will focus on **Polyline** and **Volume** support. In the next section, we will highlight **Setter** supports.

We will start demonstrating MPBF-specific supports using **Polyline** support. *Figure 12.12* shows how to apply **Polyline** support to the two faces we've selected. The **POLYLINE SUPPORT** dialogue is made up of five tabs. In the **Geometry** tab, we can select the faces we wish to support and designate an overhang angle below which Fusion 360 should create support structures:

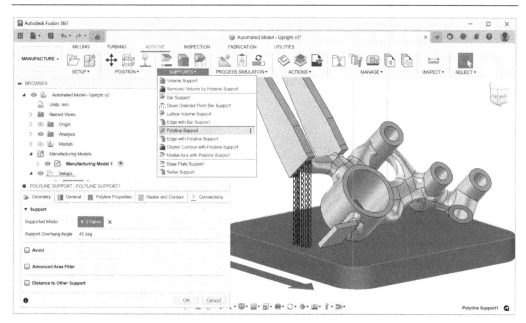

Figure 12.12 – Adding a polyline support to selected faces

The **Geometry** tab has additional functionality around how to select surfaces to avoid generating supports and how much gap there should be between each support.

If you have chosen to support an entire body or a component, by selecting the **Avoid** checkbox, as shown in *Figure 12.13*, you can select one or multiple faces to avoid creating support structures. If you activate the **Advanced Area Filter** checkbox, Fusion 360 gives you additional options around which area within your selected model/face to create support structures. If you activate the **Distance to Other Support** checkbox, you can designate a gap between multiple support structures to avoid overlap between them:

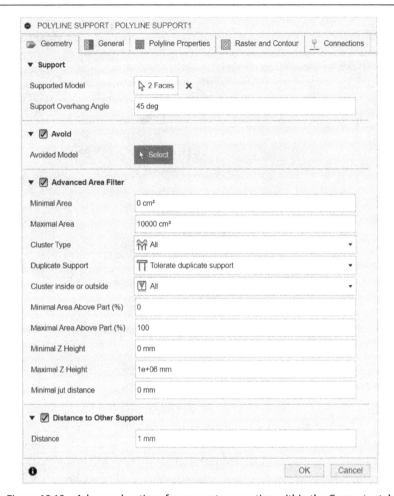

Figure 12.13 – Advanced options for support generation within the Geometry tab

The **General** tab within the **POLYLINE SUPPORT** dialogue allows you to control the polyline type with a drop-down list, as shown in *Figure 12.14*. There are three polyline types you can select. The default selection is **Structure with** hole **patterns**, as indicated with callout 1. For this polyline type, you can control **Polyline Properties** and **Connections** with their designated tabs, as shown at the top left of *Figure 12.14*. If you set the polyline type to **Surface**, as shown with callout 2 in the same figure, the **Polyline Properties** tab and the **Connections** tab will disappear. The same is also true when setting the polyline type to **Solid**, as indicated with callout 3.

The **General** tab also allows you to modify the distance between each part connection point when editing the **Anchor Distance** and **Hatch Distance** inputs. You can also control the orientation of the polyline support and whether these supports should go all the way to the edges of your part or not. By default, all polyline supports are aligned to the *Y*-axis. If you enter an input of 45 degrees in the **Rotate by Z** field, your polyline supports will be diagonal along the XY plane:

Figure 12.14 – The General tab for polyline support

If you are generating support using the polyline type of **Structure with hole patterns**, you will also be able to edit the polyline properties using the third tab within the **POLYLINE SUPPORT** dialogue, as shown in *Figure 12.15*. This tab allows you to control the hole pattern, including its density and thickness. You can also add **Fin Structures** so that you have a better connection to the build plate. By default, Fusion 360 creates each polyline support with a single hatch. This means their thickness is controlled by the melt pool diameter of the laser, as set in the print settings for the additive setup. However, you can enable **Thickening up Structures** by selecting the checkbox and assigning a specific thickness to polyline supports, making them stronger. You can control the thickness of the connection to the platform, the connection to the parts, as well as the main lattice structure:

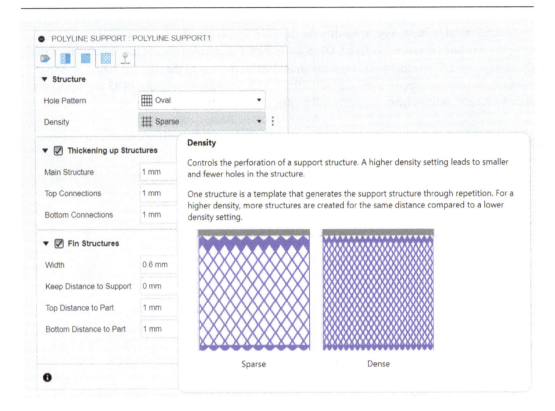

Figure 12.15 – The Polyline Properties tab for polyline supports

Within the **POLYLINE SUPPORT** dialogue, if you navigate to the **Raster and Contour** tab, as shown in *Figure 12.16*, you can control the fragment size and gap. By default, both the **Fragment Size** and **Fragment Gap** inputs are set to **Small**. If you change the input for these fields to **Custom**, you can control **Fragment Contour Length** as well as **Fragment Contour Gap**, as shown in *Figure 12.16*. **Fragment Contour Length** controls the size of a given fragment. In this example, each fragment is set to 3 mm, which means each polyline segment will be 3 mm long. **Fragment Contour Gap** is set to 0.25 mm, which means there will be a 0.25 mm distance between each polyline support segment:

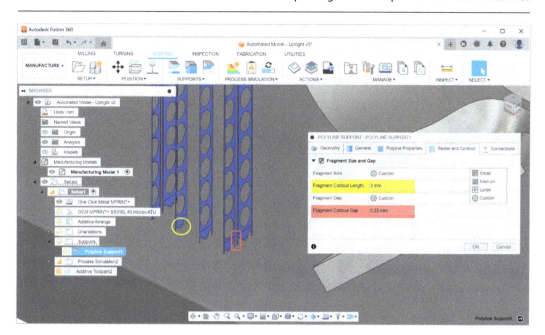

Figure 12.16 – Modifying Fragment Size and Gap for polyline support

The final tab of the **POLYLINE SUPPORT** dialogue controls the **Connections** properties of the polyline support for the part and the build plate. There are two segments to control the part connection. **Top Part Connection** controls how the polyline supports contact the downskin region of a model, while **Bottom Part Connection** controls how the polyline supports contact the upskin of a given part. **Platform Connection** controls how polyline supports are connected to the build plate. You can also turn on the **Triangular Fin on Platform** option to create additional flanges to connect your polylines to the build plate for stronger adherence.

There are four options to control the shape of a given part connection. The default connection shape is **Breaking point**. *Figure 12.17* shows the outcome of each shape. The leftmost screenshot shows the shape outcome of the **Breaking point** option. The middle screenshot shows the **Offsetted Breaking Point** option, and the screenshot on the right shows the **Strip** shape:

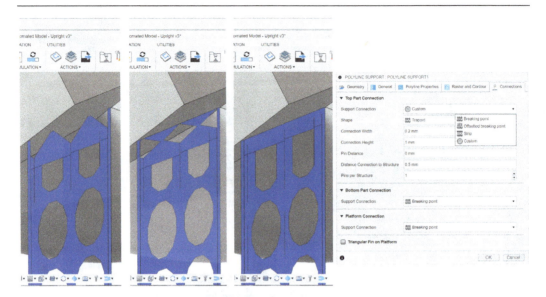

Figure 12.17 – The three shape outcomes for the top part support connection with polyline supports

The next support type we will highlight in this section is **Volume** support. You can access the **VOLUME SUPPORT** dialogue by selecting its corresponding option within the **SUPPORTS** panel of the **ADDITIVE** tab. Volume supports are very similar to polyline supports. The main difference is that volume supports create multiple polylines to create a stronger support structure. *Figure 12.18* shows the outcome of selecting a single face and creating a volume support. The same figure shows the previously created polyline supports on the left-hand side and the newly created volume supports on the right-hand side. As you can see, we can create structures with **Hole Pattern** set to **Oval** using the **Volume Properties** tab. We can also control **Fragment Size** and **Fragment Gap** so that we have large fragments of supports with contour and a raster gap of 0.25 mm to replicate the shape and size of the polyline supports we previously generated:

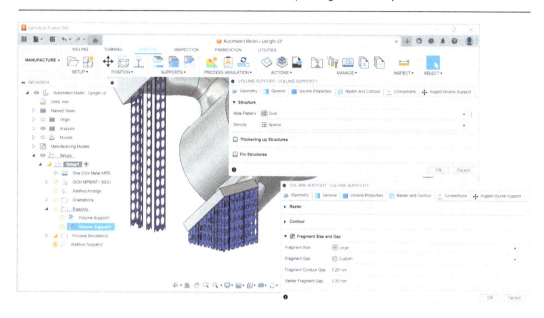

Figure 12.18 – Creating volume support for an MPBF additive setup

By default, all support structures in Fusion 360 connect selected downskin surfaces to the build plate or an upskin region of a body. *Figure 12.19* shows an updated setup, where both **Polyline Support1** and **Volume Support1**, which we created in the previous steps, are suppressed. In this figure, we can see the outcome of selecting multiple surfaces and generating a new **Volume** support with default settings. Here, we can see that the supports on the left-hand side reach down to the build plate, but the supports on the right-hand side stop on the up skin of the same part. Such support geometry is desirable if we are looking to minimize the time and material we use to manufacture our parts with support structures. But we also have to consider the additional effort we will have to exert during the post-processing phase. Specifically, any support structure that touches the upskin of a given part needs to be removed and any blemishes need to be machined away. In such cases, we may want to avoid creating support structures that touch the upskin of our parts:

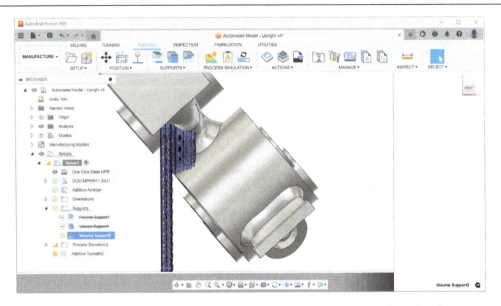

Figure 12.19 – Generating a volume support using default settings on selected surfaces

One way we can generate support that does not touch the upskin of our components is by generating angled volume supports. **Angled Volume Support** is the final tab within the volume support dialogue. Within this tab, we can activate **Angled Support** and designate the pivot point where our volume support will be broken into two segments. As shown in *Figure 12.20*, by manipulating the on-canvas gizmo, we can control the breakpoint in our volume supports so that the generated support structure is angled away from the downskin:

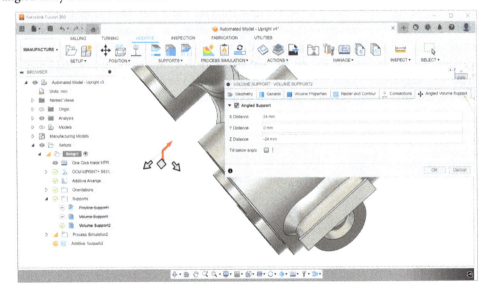

Figure 12.20 – Manipulating the pivot point for an angled volume support

By default, the angled volume support pivot point allows us to control where our support structures will pivot away from the vertical axis. As you can see in *Figure 12.21*, the bottom half of the support structure is vertical and is pointing toward the build plate. The top section of the same support structure is angled and points toward the selected downskin surfaces:

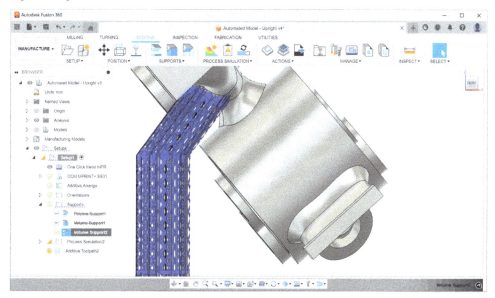

Figure 12.21 – Angled volume support

This outcome avoids creating supports that touch the up skin of our part. If you select the **Tilt below angle** checkbox within the **Angled Volume Support** tab of the dialogue, you can also control the bottom half of the support itself. As mentioned previously, the bottom half of the support is vertical by default. However, you can also control its vector by activating the **Tilt below angle** checkbox.

As was shown in *Figure 12.12*, there are multiple polyline and volume support options available for MPBF additive manufacturing setups. All these options are variations of the polyline and volume supports and can be applied to different aspects of our models, such as edges and medial lines for small area surfaces. You can also apply secondary polyline supports to surround an existing volume support. Various options to control these support structures are available within their respective dialogues.

Once we are satisfied with how our models are oriented, positioned, and supported, we can generate the additive toolpath and simulate it as we have done for FFF setups and SLA/DLP setups in the previous chapters. *Figure 12.22* shows how you can simulate the additive toolpath by selecting the appropriate command within the **ACTIONS** panel of the **ADDITIVE** tab. The **SIMULATE ADDITIVE TOOLPATH** dialogue for an MPBF setup is the same FFF and SLA/DLP setups. We can use the **Show Machine** checkbox to display or hide the machine platform. We can use the **Incremental Move** checkbox to display the layer outcome one layer at a time or in preset increments:

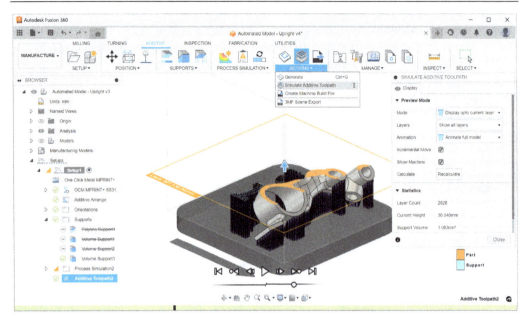

Figure 12.22 – Simulating the additive toolpath of an MPBF additive setup

Please keep in mind that MPBF toolpaths would take a longer time to calculate/recalculate if you selected the **Show all layers** option within the **Simulate Additive Toolpath** dialogue as MPBF printers have smaller layer heights.

If you just want to see a quick animation of your additive toolpath, you may want to choose to generate and show every 2nd, 5th, 10th, or 100th layer. The additive toolpath simulation shown in *Figure 12.22* was generated after suppressing the polyline support as well as the two volume supports that were used previously to demonstrate the support concepts earlier. **Volume Support3** was added to the entire component using default settings to support the entire body.

Once we are done orienting our models, adding support structures, and simulating the toolpath, we can create the necessary machine build file to start the 3D printing job. At the beginning of this section, we created the machine build file for a generic MPBF printer, as shown in *Figure 12.7*. In *Figure 12.23*, we can see that using the same **Create Machine Build File** command will result in a similar dialogue with a machine-specific export option. In this setup, we utilized a **One Click Metal** printer. Therefore, the exporter dropdown only contains the relevant file format, which is **OneClickMetal exporter (.gcode)**:

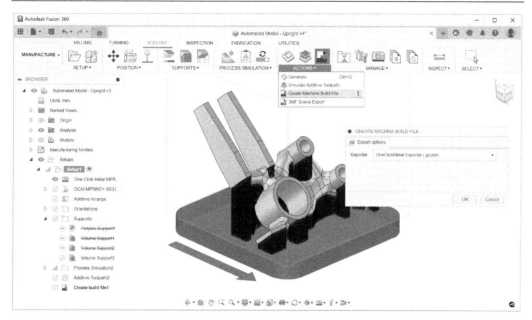

Figure 12.23 – Exporting the necessary machine build file from an additive setup

Fusion 360 can create machine build files for eight of the nine machine vendors it has in its machine library. As we just demonstrated, you can create a `gcode` file specific to an OCM machine. You can also generate the following:

- ILT files (`*.ilt`), which are required for Aconity printers

- LMI Exporter files (`*.gcode`), which are required for a Kurtz Ersa Alpha 140 printer

- Additive Industries Exporter files (`*.zip`), which are required for an Additive Industries MetalFAB1 printer

- MTT files (`*.mtt`), which are required for all Renishaw printers

- SLM files (`*.slm`), which are required for all SLM printers

- DMG Mori Lasertec files (`*.zip`), which are required for all DMG Mori printers

- Xact Metal Exporter files (`*.cli` or `*.sli`), which are required for Xact Metal printers, depending on the model number

If you are using Fusion 360 alongside your EOS Metal 3D printers, you will need to have a license to software called EOSPrint from your machine manufacturer to submit jobs to your printer. Fusion 360 can be used to create additive setups with your EOS metal printers. You can select your print settings from Fusion's print setting library, which aligns with all the print settings supplied by EOS. You can add support structures and slice your models in Fusion as well. However, to create the additive toolpath specific to your EOS hardware, you will need to have a license for EOSPrint.

If you have Fusion 360 and EOSPrint software installed and licensed on your computer, you can generate the additive toolpath and even inquire about the estimated build time for your additive setup. You can connect to your specific printer and submit a print job directly to the printer itself. To activate this functionality, you will also need to download and install *EOSPRINT for Autodesk Fusion 360*. You can locate this app on the Autodesk App Store at `https://apps.autodesk.com/` `FUSION/en/Detail/Index?id=8154614516617688086`.

In this section, we highlighted how to generate an orientation study and select a part position and orientation for a successful metal additive print using MPBF printers. We also went over how to create support structures for this additive manufacturing technology.

In the next section, we will discuss how to create setter supports, which are used during the sintering process after printing parts using metal binder jetting technologies.

Setter supports for a successful sintering process

MPBF is one of the most adopted technologies for creating metal parts using additive manufacturing. However, recently, there have been several machine vendors entering the metal binder jetting field. Desktop Metal has been a leader in this field for many years. HP also recently entered the metal binder jetting field by releasing a new printer named **Metal Jet S100**. Markforged, which is a company known for making 3D printers that can print using the continuous fiber reinforcement method, also started offering its **Digital Metal PX 100** series of printers to serve the metal binder jetting field. Alongside these well-known brands are also some new entries into the field, such as **Sinterjet M60** from Sintertek.

As manufacturers start adopting metal binder jet technology in their workflows, we need to highlight how you can create additive setups using Fusion 360 and prepare your prints for these printers. For the most part, creating an additive setup for a binder jetting technology is no different than the additive setup we create for SLS printers. The most important aspect of an additive setup for a binder jet printer is orienting and arranging our parts within the build volume. Such a setup with four components with an HP Metal Jet S100 printer is shown in *Figure 12.24*.

Some 3D printers (for example, Digital Metal PX100) require us to scale all our models up before slicing. Other printers (for example, HP Metal Jet S100) can automatically handle the scaling of parts onboard the printer.

Regardless of the metal binder jetting printer that we use for an additive setup, you may notice that the list of available support types for a binder jet setup is limited to Base Plate support and Setter support. This is because binder jetting technology does not require support structures during the printing phase. However, once the print is finished, we are left with a part in its green state. Parts printed with a binder jet printer need to be post-processed and require sintering to evaporate the strong binder holding the metal part together. This process also fully compacts the metal part to remove porosity:

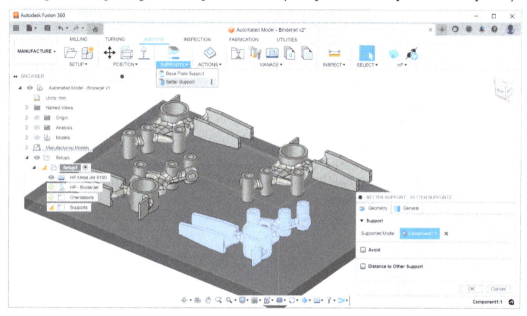

Figure 12.24 – Creating a setter support for an additive setup with a metal binder jet printer

After printing a part and extracting it from the loose powder, we will need to place it in a sintering oven. Parts that just finished curing in the printer are called green parts. In this state, they are fragile and need to be held in place with a material that can both cradle the part and shrink with it during the sintering process.

Using the **Setter Support** command located within the **SUPPORTS** panel of the **ADDITIVE** tab, we can select a given component to generate a support structure, which will primarily be used during the sintering process.

Figure 12.25 shows the **General** tab for the **SETTER SUPPORT** dialogue. In this tab, we can control various options to create a setter support to protect our parts during the sintering process. The **Gap** input controls the distance between the part and the setter support. The **Z Overlap** drop-down menu is set to **Remove** by default, which allows a part to be lifted in the Z direction without any support structure blocking the motion. **Bounding Shape** is the **Silhouette** component of our parts by default, which generates a shape based on its shadow. The **Fill Inner Contour** option is set to **Yes** by default, which creates a support structure, even if there is a hole aligned with the Z axis of our part. The **Outer Offset (XY Plane)** input allows us to control the thickness of the setter around the perimeter of the part. The **Maximum Height** input allows us to control how high our setter supports should be from the build plate. The **Taper Angle** input controls the angle from the vertical axis for the perimeter of our setter support perimeter walls.

As we are choosing an orientation and position for our parts within a binder jet machine, it is important to place our parts above the Z-axis by a certain amount so that the setter supports we generate can cradle the parts along its bottom surface as well as its side surfaces.

Figure 12.25 shows a close-up view of our additive setup while hiding **Component 1:2**, **Component 1:4**, and **Component 1:5** and only displaying **Component1:1**. To improve the visualization, we also hid the machine build plate in this figure:

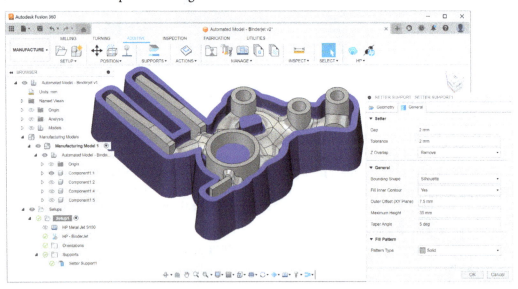

Figure 12.25 – Generating a setter support for a component

Another option we have with setter supports is to modify the bounding shape from a silhouette of our parts to a rectangular object. As shown in *Figure 12.26*, we can set **Bounding Shape** to **Rectangle** and change the pattern type from a **Solid** pattern to a **Rectangular**, **Hexagonal**, **Circular**, or **Column** pattern. *Figure 12.26* shows the outcome of a hexagonal pattern. If we choose a pattern other than solid, we can also control the pattern size by adjusting its cell size and wall thickness:

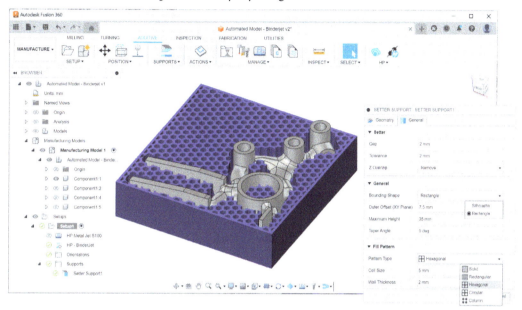

Figure 12.26 – A setter support with a rectangular bounding shape and a hexagonal fill pattern

Once we are satisfied with our support, we will need to split the support structure from the part as we may not want to 3D print them in place. Please remember that we don't need the support structure during the printing process. We will only utilize it during the sintering phase. Therefore, we don't need to print the part and the support structure together. We can accomplish this by exporting our additive setup as a 3MF file using the **3MF Scene Export** function located in the **ACTIONS** panel of the **ADDITIVE** tab, as shown in *Figure 12.27*. When we use this command, we will be able to modify the options within the **3D SCENE EXPORT** dialogue and select the **Include the support structures** checkbox:

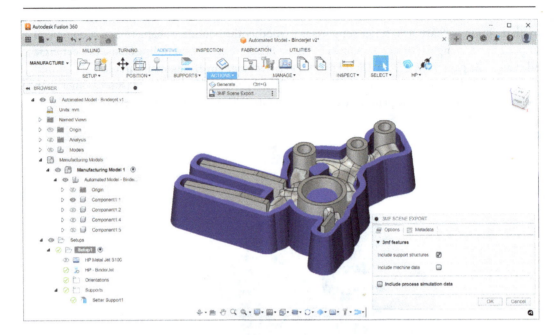

Figure 12.27 – Creating a 3MF file and including support structures

Once the 3MF file has been saved on our hard disk, we can create a new Fusion document by selecting the + icon, as shown in *Figure 12.28* via callout 1. Within the **MESH** tab of the **DESIGN** workspace, we can use the **Insert Mesh** function, as shown in callout 2, to insert the 3MF file we have just created into the active design document. Fusion will import each component as a mesh body with a unique name. We can simply hide the visualization for mesh bodies 3, 4, and 5, as shown by callout 3. This way, we will be able to focus on the support structure and mesh body of the component we created the setter support for:

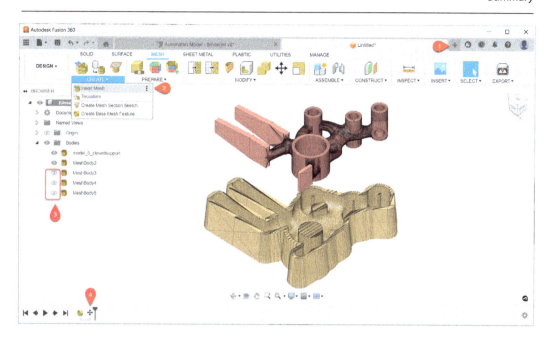

Figure 12.28 – Inserting a 3MF file into a Fusion document to visualize setter support outcomes

Using the **Move** command located in the **Modify** tab, you can select **Meshbody2** and move it to the Z-axis. *Figure 12.28* shows the outcome of this move operation, as highlighted by callout 4 in the timeline. As you can see, our bracket component fits neatly within the setter support and can be removed from the Z-axis without any blocker. If we put this part and the setter support into the sintering oven together, the binder will evaporate from the part and the support and these parts will shrink together uniformly. The setter support will support the component during this sintering process so that it does not sag or permanently deform during the sintering phase.

Summary

In this chapter, we learned all about metal additive manufacturing. We started this chapter by highlighting all the additive manufacturing technologies you can utilize to create metal parts. We demonstrated how to use fused filament fabrication printers from UltiMaker and Raise3D and metal filaments from companies such as BASF to 3D print metal parts that require debinding and sintering. We also showcased how you can use any FFF 3D printer with customized extruders and specialized metal filaments from companies such as The Virtual Foundry to 3D print with metal filaments. Each of these solutions requires the 3D-printed part to be further processed by debinding and sintering. We then highlighted how to create metal parts using directed energy deposition technology utilizing printers such as the Meltio M450 printer available within Fusion 360's machine library.

In the next section, we demonstrated how to create additive setups using MPBF printers. We showcased how additive setups for MPBF printers are very similar to the setups we created for FFF and SLA/DLP 3D printers, with one main difference around part positioning and support structure generation. We highlighted how to stagger part positions so that we do not create a high resistance zone for a hard recoater blade. We also highlighted how to create various polyline and volume supports to support the part during the printing process. We showcased how to minimize material use to create those support structures and avoid support creation on upskin surfaces by utilizing angled volume supports. At the end of the second section, we also covered how to generate the machine-specific file format needed to initiate a print job after slicing and simulating our additive toolpath.

In the final section, we focused our efforts on the metal binder jetting process. We discussed how to create an additive setup for this process. We highlighted how the loose powder within the binder jetting technology is capable of supporting parts being printed during the curing phase. We also discussed how parts manufactured with this technology require a sintering process to evaporate the binding agent and result in a fully dense part. In the final section of this chapter, we covered how to create setter supports that can cradle our parts and protect them from sagging during the sintering process while shrinking uniformly with our part as they are made with the same process and material.

Using any one of these techniques, you can feel more confident in creating your additive setups for the 3D printing technology you have available to generate a metal component. In the next chapter, we will focus on how to simulate the MPBF process to identify and fix potential print failures that are caused due to high thermal stresses and distortions that are common with the process.

13

Simulating the MPBF Process

Welcome to *Chapter 13*. In this chapter, we will explore how to simulate the printing process within a **metal powder bed fusion** (**MPBF**) 3D printer. MPBF machines are among the most expensive 3D printers. Metal powder is an expensive form of 3D printing material. Therefore, any mistake or print failure is a costly one when it comes to metal 3D printing. Fusion 360 users with Manufacturing extension access can simulate their MPBF printing process to detect and rectify common print failure modes, such as part distortion and recoater blade interference. In this chapter, we will highlight how to perform such analyses and make the necessary changes to avoid common print failure modes.

In this chapter, we will cover the following topics:

- Setting up a process simulation
- Interpreting the results
- Compensating for distortions and updating the setup

By the end of this chapter, you will have learned how to select a component in an additive setup with an MPBF printer as a target for process simulation. You will acquire an understanding of the various print settings and how they impact process simulation inputs. You will learn how to modify process inputs, such as print bed heating, and finite element analysis inputs, such as mesh settings to conduct a process simulation, while making trade-offs of speed versus accuracy. You will also learn how to inspect the results of process simulation and identify large deformation areas in your model, determining whether your parts will experience a recoater crash or not. By the end of the chapter, you will be able to export a compensated shape to offset the distortions your parts may experience during the printing process, printing your parts with minimal distortions.

Technical requirements

In this chapter, we will continue to cover metal additive manufacturing. We will create additive setups using an MPBF 3D printer and its respective print settings. We will generate support structures specific to MPBF machines. We will also conduct process simulation for our parts and supports. MPBF machines, MPBF-specific support structures, and the ability to conduct a process simulation are not available with Fusion 360 until you gain access to the Manufacturing extension.

As with all other Fusion 360 extensions, the Manufacturing extension is not available for personal use.

If you have a commercial license of Fusion 360, to create an additive setup with an MPBF machine, you will also need access to the Manufacturing extension (`https://www.autodesk.com/products/Fusion-360/manufacturing-extension`).

Autodesk offers a 14-day trial for the Manufacturing extension for all commercial users of Fusion 360. You can activate your trial within the extensions dialog of Fusion 360.

The Manufacturing extension is also available for free to Startup and Educational license users.

The lesson files for this chapter can be found here: `https://github.com/PacktPublishing/3D-Printing-with-Fusion-360`

Setting up a process simulation

In this section, we will demonstrate how to create a metal additive manufacturing setup using a powder bed fusion printer and initiate a process simulation, in order to analyze the deformation a part will experience during the 3D-printing process.

In order to demonstrate this functionality, we will use the Fusion document named `L Brackets`, as shown in *Figure 13.1*. This model consists of three components with varying overhang angles. `Component1` has an overhang angle of 0 degrees, `Component2` has an overhang angle of 15 degrees, and `Component3` has an overhang angle of 30 degrees.

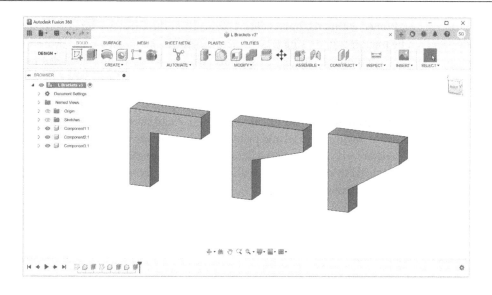

Figure 13.1 – Three components with various overhang angles

After changing the Fusion workspace from **DESIGN** to **MANUFACTURE**, we can create an additive setup using a **Renishaw AM 250** machine from Fusion's machine library. In this setup, we will select the **RenAM250 Titanium Ti6Al4V-1123-Q 60 micron** print setting, as shown in *Figure 13.2*. We will also manually arrange all three components such that they are staggered in the X and Y directions, ensuring that they do not interfere with the recoater blade direction, as indicated on the build plate. As you may recall, in *Chapter 12*, we discussed the reasoning behind why such an arrangement is beneficial when preparing our builds for printing with an MPBF machine.

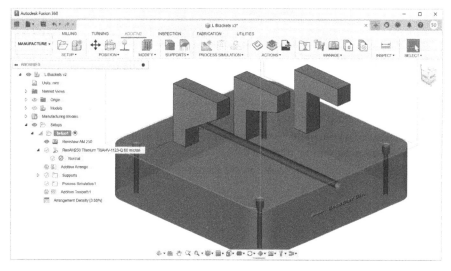

Figure 13.2 – Arranging three components within an additive setup

Once our parts are positioned within the build volume, we can add the necessary support structures so that our parts can be manufactured using this printer. As you can see in *Figure 13.3*, we added a volume support for each component. During the support creation, we used the default settings for the volume support.

As a reminder, the default volume supports are generated with a diamond hole pattern, using a sparse density input. **Thickening up Structures** and **Fin Structures** are enabled, which creates a supporting geometry as thin as possible, saving material and print time. However, we do not know whether the support structures we generate will be strong enough to support the part during the printing process. That is another reason we need to conduct a process simulation – to see whether we need to make support stronger to avoid a potential print failure.

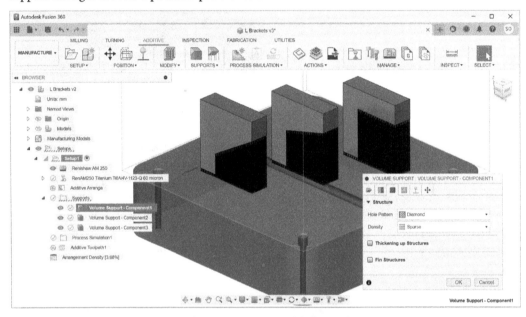

Figure 13.3 – Volume supports are added to each component

Before we initiate a process simulation, it is a good idea to take a closer look at our print settings. If we select the print setting associated with this setup, right-click, and edit it, we will see the **Print Setting Editor** dialog and can inspect the build parameters included in this print setting. As highlighted in yellow in *Figure 13.4*, this print setting utilizes a layer thickness of 60 microns, and the material is set to titanium. Both of these inputs can be found in the **General** section of the print setting.

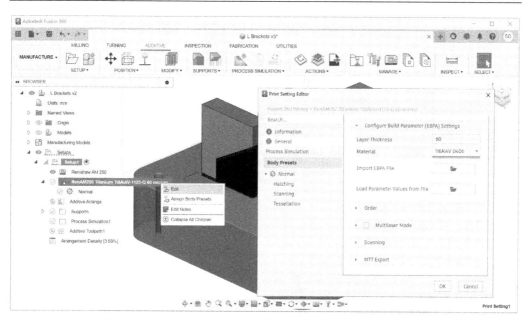

Figure 13.4 – Editing the print setting to visualize build parameters

If we select the **Hatching** subheading within the **Normal** body preset, we will also be able to visualize the hatch distance and rotation increment angle for the hatch pattern that Fusion 360 will utilize to create the infill for this print. As shown on the left side of *Figure 13.5*, the current strategy for the infill is the **Stripe** method. Each laser hatch for this pattern will be 0.095 millimeters apart, and the hatch pattern will alternate 67 degrees around the *Z* axis per layer.

If we select the **Scanning** subheading within the **Body Presets** section for the **Normal** preset, we will also be able to visualize the laser power associated with this print setting. The image on the right side of *Figure 13.5* shows that the laser will be set to 200 watts while printing the hatches.

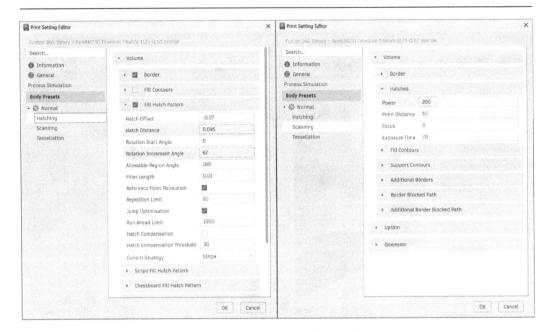

Figure 13.5 – Hatching and scanning print settings for titanium

The information we have highlighted in *Figure 13.3* and *Figure 13.4* are parameters recommended by the machine manufacturer in order to print with titanium on this specific printer. As you may have noticed, there are many parameters you can edit to fine-tune your print settings for an MPBF printer. Each parameter has an impact on the outcome of your print.

However, only a subset of these parameters are taken into consideration when conducting a process simulation, as they have been identified as having the most significant impact on thermally induced stresses during the printing process. The parameters we have highlighted in the previous figures feed directly into the inputs required to conduct a process simulation in Fusion 360.

As you can see in *Figure 13.6*, if we select the **Process Simulation** subheading within **Print Settings Editor** for this print setting, we can see that Fusion 360 already has an associated parameter set to utilize for the simulation. Process simulation parameter sets are saved as ∗.PRM files.

> **Important note**
> "A **processing parameter** (**PRM**) file records machine process parameters and material properties. A moving heat source is applied to the material on a small-scale analysis. The PRM file stores the mechanical response of the material to the machine parameter setting. This can be extrapolated to a full part … the PRM file can be reused for any future simulation using the same combination of material and processing parameters" (https://help.autodesk.com/view/NETF/2024/ENU/?guid=GUID-7A9E551D-36B2-4922-8A38-0072EEDB5A23).

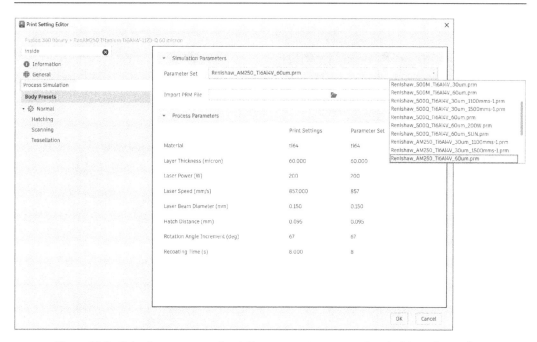

Figure 13.6 – Selecting a process simulation parameter set associated with a print setting

If you want to conduct a process simulation using a different PRM file, you can do so by selecting it from the drop-down menu named **Parameter Set**. This dropdown automatically filters simulation parameters within its library to PRM files associated with the same material selected within the print setting. As shown in *Figure 13.7*, after changing the parameter set from the default PRM of **Renishaw_AM250_Ti6Al4V_60um** to **Renishaw_AM250_Ti6Al4V_30um_1500mm/s-1**, **Print Setting Editor** indicates that the additive toolpath will be created using the original print setting, yet the process simulation will be based on the newly selected parameter set. In the same figure, you can see the highlighted values, which show the difference between the **Renishaw_AM250_Ti6Al4V_60um** print setting versus the **Renishaw_AM250_Ti6Al4V_30um_1500mm/s-1** parameter set inputs for **Layer Thickness** (60.000 microns versus 30.000 microns) as well as **Laser Speed** (857.000 mm/s versus 1500 mm/s) and **Hatch Distance** (0.095 mm versus 0.065 mm).

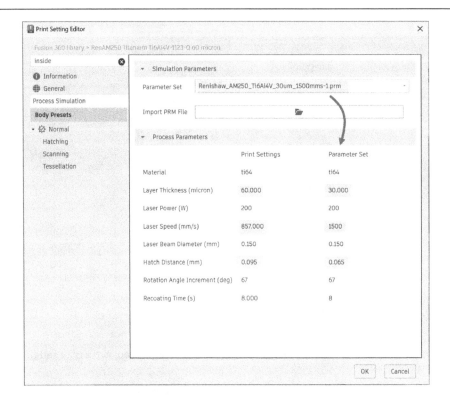

Figure 13.7 – The differences in process parameters between the print setting versus the non-default PRM

Now that we have highlighted how Fusion 360 utilizes the inputs we have in our print settings to conduct process simulation, we can close the **Print Setting Editor** dialog and proceed with the default print settings. In order to start setting up our simulation, we need to access the **Study** command located within the **PROCESS SIMULATION** panel of the **ADDITIVE** tab. As shown in *Figure 13.8*, this will bring up the **STUDY** dialog, where we can select the target body we wish to conduct our process simulation on. The **Advanced Settings** section within this dialog allows us to control the size of the mesh elements that Fusion 360 will create to conduct this finite element analysis. The **Layers per element** field controls the size of a given element, based on the layer thickness that was set within the print settings. By default, Fusion 360 utilizes an adaptive mesh, where the elements are smaller at the layer that is being printed and grow larger below the active layer. An adaptive mesh results in a fast analysis at the cost of accuracy. The **Coarsening Generations** and **Max Adaptivity Level** inputs control the change in element size below the layer being printed. Using smaller inputs for these two fields will increase the accuracy of your results at the cost of increased solve times. You can also control the **Platform Thickness** and **Platform Heating** inputs within this dialog.

In order to simplify the meshing process and solve times, Fusion 360 homogenizes the support structure geometry and simulates it using modified physical properties. When the **Auto Calculate Support Volume Fraction** checkbox is active, this homogenization is performed automatically.

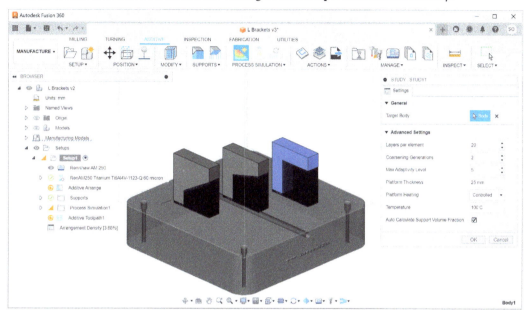

Figure 13.8 – Setting up a process simulation using the STUDY dialog

After selecting our target body and making all the customizations necessary within the advanced settings of the **STUDY** dialog, we can select **OK** to proceed. At this point, Fusion 360 conducts a pre-check and displays the **SOLVE** dialog.

In *Figure 13.9*, we can see that our print settings match our PRM file, which means we can use the default parameter set to solve this simulation and do not need to select a different parameter set to proceed.

As we chose to utilize the default input values in the **STUDY** dialog for **Layers per element** (**20**), **Coarsening Generations** (**2**), and **Max Adaptivity Level** (**5**), our solve accuracy will be low, but we will calculate our results quickly.

Our model has the appropriate contact to the platform and does not require an additional support structure.

We have not generated the mesh at this point; therefore, we see a warning sign next to the *FEA mesh* section. We can either select the **Check mesh** icon or simply select the **Solve** button to start the simulation.

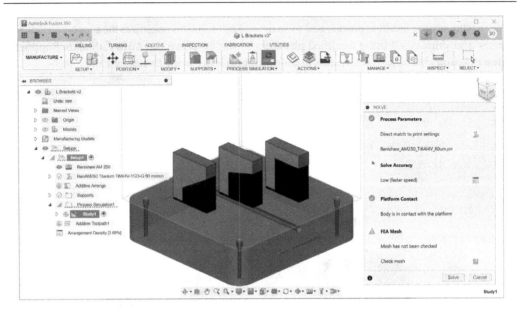

Figure 13.9 – A pre-check before starting a process simulation

Once a solve is initiated, Fusion 360 displays the **CAM Task Manager** dialog, as shown in *Figure 13.10*. In this dialog, we can see that Fusion conducts a thermo-mechanical simulation to calculate the deformations and stresses that our part will be subjected to during the 3D printing process. If we select the line item for the active study, we can right-click and disable the **Auto Remove** option, allowing us to monitor the progress during the solve and inquire the total solve duration once it is complete.

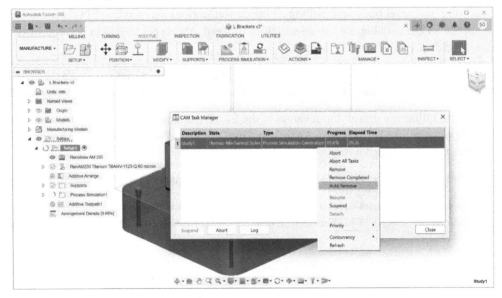

Figure 13.10 – CAM Task Manager reports the progress of a process simulation

After the thermo-mechanical simulation is complete, Fusion 360 will display the results of the simulation in a new tab called **RESULTS**, as shown in *Figure 13.11*. We will see a colorful representation of the build plate and the first few layers of the part being 3D-printed. As this is a transient analysis, we will be able to toggle through the time steps to visualize the part being printed, layer group after layer group.

Figure 13.11 – Visualizing the results of the process simulation

The default colors we will see represent the magnitude of the displacement our part will experience during the printing process, and they will correspond to the values shown in the legend on the right side of the canvas.

In this section, we have highlighted how to create an additive manufacturing setup with an MPBF printer. After orienting and supporting our components, we initiated a process simulation. We covered how to inquire about the print settings and how they are associated with the process simulation parameters. We selected a component and conducted a process simulation using it.

In the next section, we will learn how to review the results of the process simulation and export the deformed and compensated versions of the component we simulated as STL or 3MF files.

Interpreting the results

We ended the previous section by completing our first process simulation. In this section, we will focus on the various types of results Fusion 360 calculates when conducting a process simulation.

As you can see in *Figure 13.12*, Fusion calculates and displays six result types for an MPBF process simulation. Once the analysis is done, the default result type we see is **Displacement**. However, I recommend taking a look at the **Structure Type** result after generating a mesh or completing the process simulation. This result type colors each element based on whether it represents a support structure, a part, or the build plate.

Segments of the part that have a down skin without a support structure are colored red to represent under-constrained zones. Fusion also colors support structures that are discontinuous and the *Z* axis with unintended gaps as under-constrained elements. In such cases, you may want to go back to the process simulation setup and edit your mesh settings, in order to properly mesh your support structures or generate additional support structures before meshing/solving your model.

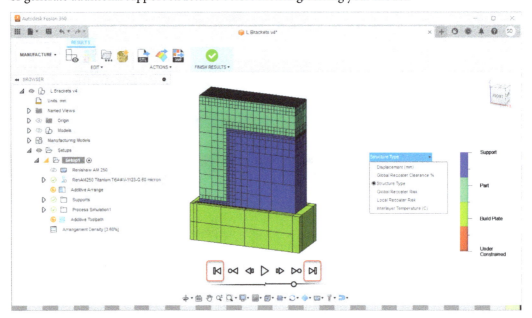

Figure 13.12 – A structure type result for a process simulation

After observing the structure type result, by using the dropdown next to the legend, you can switch back to the **Displacement** result. **Displacement** results can be further filtered down such that you can observe the deformation of your model in the **X**, **Y**, and **Z** axes. Alternatively, you can look at the magnitude of displacement. The legend and the color contour applied to the elements can be modified so that they are set to the global displacement result, or the local time step you're observing. In *Figure 13.13*, we can see the displacement in **Z** at the final time step. You can switch the time step using the simulation player toggles, as highlighted in *Figure 13.12*.

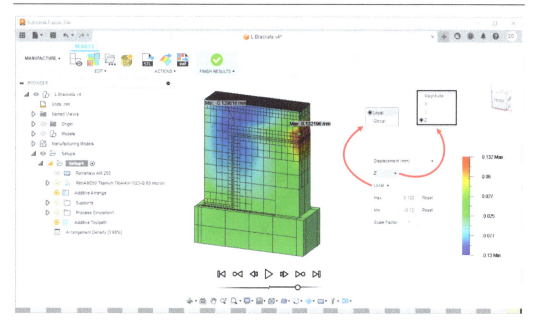

Figure 13.13 – The Displacement result type for a process simulation

Alternatively, you can use the timeline at the bottom of the screen and drag the black vertical bar back and forth to switch to a different time step. *Figure 13.14* shows the **Z** displacement result of this simulation two steps from the last time step, as highlighted by callout **1**.

Process simulation results can also be scaled to exaggerate the deformed shape by a factor, as highlighted by callout **2**. In the following screenshot, the scale factor is set to 20, which amplifies the displacement in order to show the direction of the deformation. The **EDIT** panel within the **RESULTS** tab lets us toggle element edges, allowing us to observe the results without the black lines displayed at the edge of each element. Using the same panel, you can also select **Toggle Min/Max Probes** so that you can be informed about the location of the node with the largest and smallest displacement at each time step. You can also use the **Open Results Folder** command to locate where all the simulation inputs and outputs are stored on your computer. This will also allow you to delete simulation result files if you need to clean up some space on your hard disk, as process simulation results can take up a lot of hard disk space.

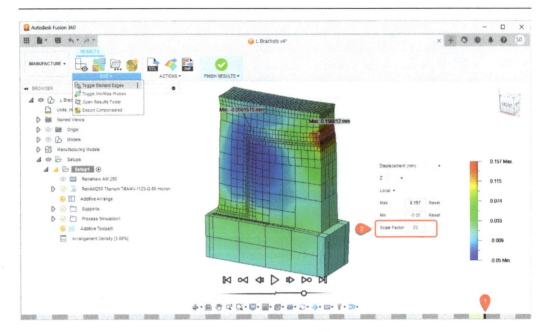

Figure 13.14 – Switching time steps and modifying the scale factor

The next result type we will highlight is the recoater clearance. During the printing process, as the laser moves over the powder, the metal powder melts and then solidifies. As the metal heats and cools during this cycle, it experiences thermally induced stresses. During the printing process, the layer that was just printed will experience deformations in the X, Y, and Z directions. If a given layer deforms in the positive Z axis an amount above the layer thickness we set within our print settings, our part will experience a recoater crash. Fusion 360 can predict such phenomena using process simulation. The recoater clearance percentage result type allows us to visualize how close our part is to the recoater blade during the printing process. A 100% clearance means our part is not experiencing any Z deformation at a given point. A 0% clearance means that our part is experiencing a Z deformation equivalent to the layer thickness. Fusion 360 displays the summary of this result for all layers in a single result type called **Global Recoater Clearance**, as shown in *Figure 13.15*. A clearance value above 100% refers to a given point experiencing a negative Z-direction deformation.

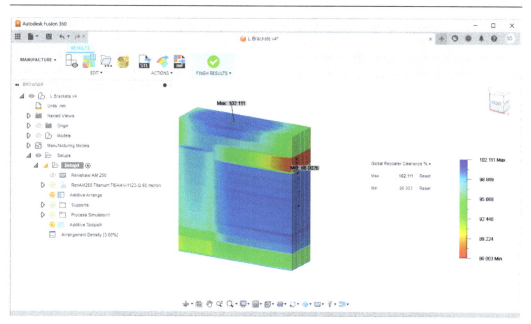

Figure 13.15 – A Global recoater Clearance % result for process simulation

As mentioned previously, Fusion calculates the recoater clearance percentage for each time step. You can visualize a simplified version of these results at each time step by switching your result type to **Local Recoater Risk**. This result type presents the recoater clearance percentage using a legend resembling a traffic light color contour of red, yellow, and green. Any recoater clearance above 80% is represented as **Low** risk for a recoater crash. If the recoater clearance is below 80% but above 40%, Fusion will highlight those elements as **Medium** risk for a recoater crash. Any clearance below 40% will be highlighted as **High** risk for a recoater crash and will be colored in red.

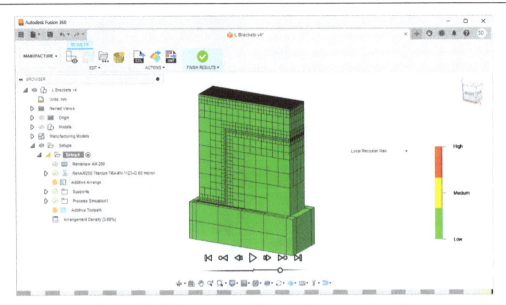

Figure 13.16 – The Local recoater Risk result for a process simulation

Fusion 360 also allows us to export the process simulation results as mesh files, as shown in *Figure 13.17*. By selecting the **Export Color 3MF** command within the **ACTIONS** panel of the **RESULTS** tab, we can extract a colorful mesh file of our results. A 3MF file can be utilized when creating simulation reports, using Microsoft Office tools such as Word or PowerPoint.

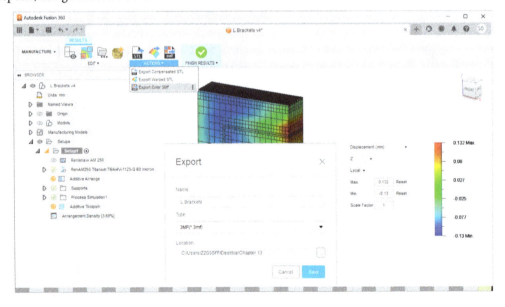

Figure 13.17 – Exporting the results of a process simulation as a 3MF file with colors

When we are done interrogating the results of the process simulation, we can simply select the **FINISH RESULTS** command from the **RESULTS** tab, which will close the results visualization and take us back to the **ADDITIVE** tab of the **MANUFACTURE** workspace.

Later on, if we wanted to interrogate the results of a given study, we could select the study from the browser, as shown in *Figure 13.18*, and select the **View Results** command located within the **PROCESS SIMULATION** tab, as highlighted with callout **1**. Alternatively, we could right-click on the selected study within the browser and select the **View Results** command from the menu, as highlighted by callout **2**.

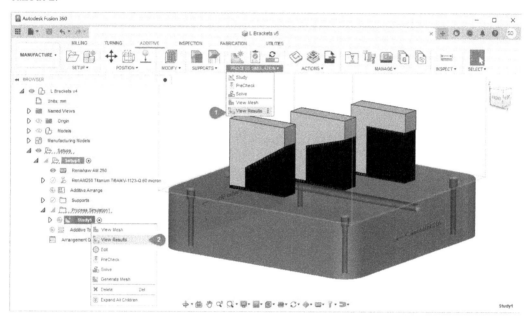

Figure 13.18 – Accessing the results of a process simulation

This action will take us back to the results visualization for the active process simulation study. Using the **RESULTS** tab, we can continue to probe our results or export a 3MF file to generate a report of our results.

In this section, we highlighted how to utilize the **RESULTS** tab to inquire about the different result types for a process simulation. In the next section, we will explore how to generate a compensated model based on the process simulation results, in order to manufacture our parts with minimum deformations.

Compensating for distortions and updating the setup

We ended the previous section by exporting a 3MF file with colors so that we could generate interactive reports with Microsoft Word or PowerPoint. In this section, we will highlight how to export warped and compensated STL files out of our process simulation results, and how to create a new setup using the compensated shape.

To get started, let's go back to the process simulation results using the **View Results** command we covered in the last section. Once we are in the **RESULTS** tab, we can select **Export Compensated STL** and **Export Warped STL** within the **ACTIONS** panel, as shown in *Figure 13.19*.

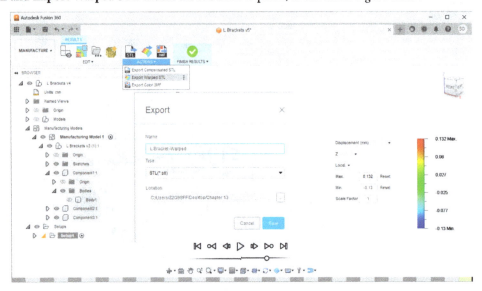

Figure 13.19 – Exporting the warped shape as an STL file

Once we create the warped and compensated STL files, we can compare them using Fusion 360. As shown in *Figure 13.20*, we can create a new Fusion design document, which takes us back to the **DESIGN** workspace. Within this workspace, we can go to the **MESH** tab and insert both of these STL files into the same document. As shown in *Figure 13.20*, this will leave us with two mesh bodies in the browser, named `L Bracket - Warped` and `L Bracket - Compensated`. If we zoom in to the top-left corner while looking at the front view of the view cube, we can see that there is a visible difference between the warped shape and the compensated shape. If we were to print our part using the as-designed shape with the supports we have generated, the outcome from the printer would be close to the shape represented by the warped object, which is highlighted by the green-shaded mesh body in the following figure. Note that in this corner of our component, the warped mesh body deforms toward the positive X direction. The compensated shape Fusion 360 generated offsets for that distortion by creating a geometry that leans toward the negative X direction.

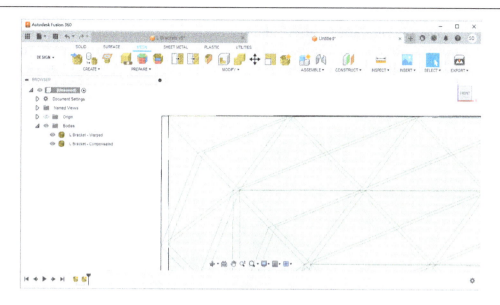

Figure 13.20 – A superimposed view of the warped and compensated models

We could also use the warped STL output in order to compare the simulation results to the printed distortions if we benchmarked the accuracy of the simulation results that Fusion 360 generates. After 3D printing our model, we can inspect the deformations using blue light scanning and compare the distorted shape we printed to the warped output that Fusion 360 predicted, checking the accuracy of the results we get using the PRM file we selected within the print settings for this study.

The workflow I recommend using after conducting a process simulation is to select the **Export Compensated** command, located within the **EDIT** panel of the **RESULTS** tab, as shown in *Figure 13.21*.

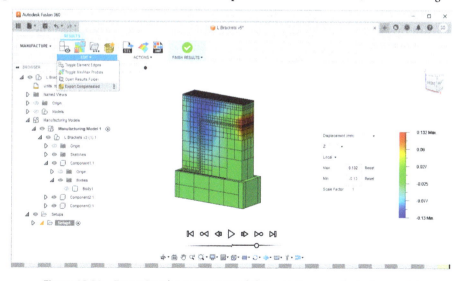

Figure 13.21 – Exporting the compensated shape to the manufacturing model

Using this function, you can generate a mesh body of the compensated model and include it directly within the manufacturing model that was used to create the active additive setup, as shown in *Figure 13.22*. The same figure displays components 1, 2, and 3, using 50% opacity so that we can clearly see the new mesh body without any obstruction.

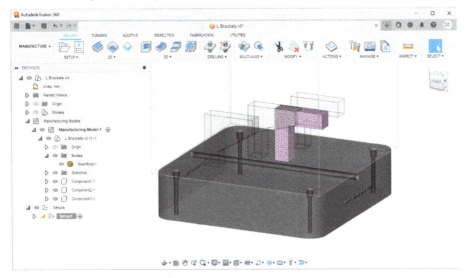

Figure 13.22 – The compensated shape included in the manufacturing model

At this point, we still need to do a few manual modifications to the manufacturing model so that we can create a setup using the compensated mesh body. First, we will select the manufacturing model, right-click, and edit it. Next, we will have to create a new component from the mesh body by right-clicking on it and selecting the **Create Components from Bodies** command, as shown in *Figure 13.23*. This action will generate a new component named Component 4, which includes the compensated mesh body. Now that we are done modifying the manufacturing model, we can select the **FINISH EDIT** command in the ribbon.

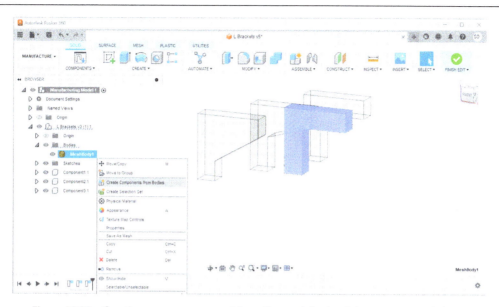

Figure 13.23 – Creating a new component from the mesh body of the compensated shape

In order to preserve our work in Setup1, we may want to duplicate it and create Setup1 (2), as shown in *Figure 13.24*. This way, we maintain our process simulation inputs and results that led us to the compensated shape. In the future, if we want to review our results or try conducting different simulations, we can simply use Setup1, which does not consider the newly created Component4.

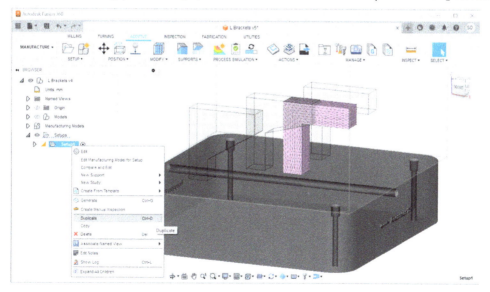

Figure 13.24 – Duplicating Setup1 to create Setup2

Now that we have a new setup, we need to edit it and deselect components 1, 2, and 3 so that we can select `Component4`, as shown in *Figure 13.25*, in order to position and support it. We can accomplish this by right-clicking on `Setup1 (2)` in the browser and selecting **Edit** to access the **SETUP** dialog.

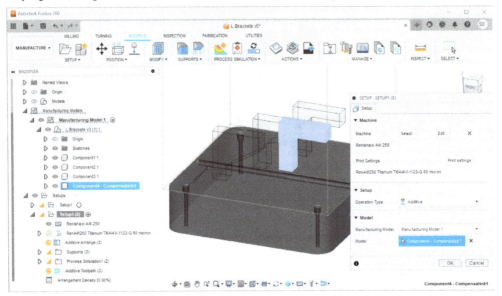

Figure 13.25 – Editing Setup1 (2) to select the compensated component

Now that we have chosen `Component4` as the only component in `Setup1 (2)`, we will have to go through the typical print preparation steps we have learned throughout this book. We can start by placing our parts within the build volume of our printer, using the **Arrange** command located in the **POSITION** panel of the **ADDITIVE** tab. *Figure 13.26* shows how to arrange the platform clearance of 3 millimeters, placing the part within the build area of our printer.

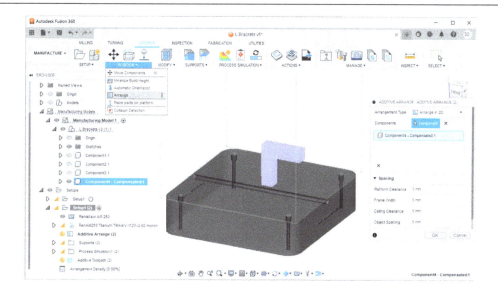

Figure 13.26 – Arranging the compensated part within the build volume

The next item on our list is to create support structures for this component. *Figure 13.27* shows how we can add volume supports to this component by selecting the appropriate support action within the **SUPPORTS** panel of the **ADDITIVE** tab. As we duplicated Setup1 to create Setup1 (2) early on in the process, all the volume supports that were used in Setup1 are still available in Setup1 (2). We need to either select and delete them or suppress them, as they will not participate in this setup.

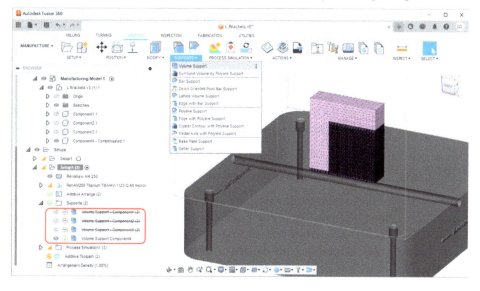

Figure 13.27 – Applying volume support to the compensated model

Once we have the compensated model positioned and properly supported, we can generate the additive toolpath and simulate it, as shown in *Figure 13.28*. Please remember that in order to generate the toolpath, you have to select the **Additive Toolpath** line item from the browser and use the **Generate** command within the **ACTIONS** panel of the **ADDITIVE** tab. After visualizing the additive toolpath layer by layer, if we are satisfied with the outcome, we can select **Create Machine Build File**, as shown in the following figure.

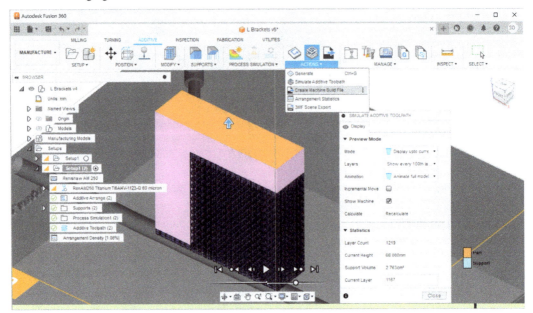

Figure 13.28 – Simulating the additive toolpath and creating the machine build file

In this example, generating the machine build file will allow us to create the necessary MTT file for a Renishaw printer. As this file is generated using the compensated model, which was created after conducting a process simulation, the resulting 3D-printed part will have negligible deformation when compared to the as-designed CAD model.

Summary

In this chapter, we focused on how to create a setup for an MPBF 3D printer and how to simulate the process using finite element analysis. The MPBF process is one of the most expensive 3D-printing technologies, both in capital expenditure and material cost. Process simulation helps us identify whether our additive setup will produce the desired outcome or not. If we are able to detect and remedy potential failures such as recoater blade crashes or large distortions before 3D printing, we can save a lot of time and money during the manufacturing phase.

In the first section of this chapter, we focused on how to generate an additive setup with an MPBF machine and highlighted the various parameters within the print settings that control the process simulation input. We also showcased how to set up a process simulation study. We demonstrated how to define process conditions, such as build plate heating, and modify key inputs to the finite element analysis solver, such as mesh settings. We ended the section by conducting our first process simulation study.

Then, we focused on the result types Fusion generates for an MPBF process simulation. We highlighted how to switch between various result types, such as structure type, displacement, recoater blade clearance, and recoater blade risk. We ended the section by exporting our results as a 3MF file with color data to be used to create reports.

In the final section, we focused on how to export the warped and compensated models from the process simulation results. We showcased how to reimport the warped and compensated models back into Fusion to visualize the difference between them. We ended the chapter by demonstrating how to create a new setup using the compensated model, in order to 3D-print a part that will deform into the as-designed model.

Using the practical knowledge you have gained in this chapter, you can feel more confident in simulating a metal additive manufacturing setup using Fusion 360, in order to identify potential print failures and compensate for distortions that are inherent in the MPBF printing process, achieving successful prints.

In the next chapter, we will focus on how to automate repetitive tasks such as part orientation, part arrangement within a build volume, and support structure generation using presets, templates, and scripts.

14
Automating Repetitive Tasks

Welcome to *Chapter 14*. In previous chapters, we demonstrated how to use Fusion 360 to design new models and import models from external sources. We also highlighted how to generate a setup to additively manufacture those models with various 3D-printing technologies. Now that we have a good understanding of how to use Fusion 360 and have experience in creating setups, arranging our parts, and creating support structures for various additive technologies, let's learn more about how to automate and optimize Fusion 360, so that we don't have to edit every input dialog for every single additive setup.

In this chapter, we will highlight how to automate various aspects of Fusion in order to minimize our interaction with the software, creating an additive setup and generating a toolpath. We will start the chapter by highlighting existing automations within Fusion's ecosystem by introducing you to the Autodesk App Store. Then, we will highlight how to customize Fusion's machine and print setting libraries, in order to create a fully defined machine that you can utilize when creating your setups. You will then learn how to customize your inputs for various operations, such as automatic orientation studies, part arrangement, and support structure generation. We will also demonstrate how to save those inputs as user defaults. Then, we will create and save templates, which combine multiple operations into a single file that can be reused on subsequent additive setups. We will end the chapter by highlighting how to utilize Fusion's programming capabilities, by creating a Python script to automate the entire process of generating an additive setup, orienting parts, arranging them within the build plate, and simulating the additive toolpath.

In this chapter, we will cover the following topics:

- Using apps from the Autodesk App Store
- Customizing presets and using templates
- Automating the additive workflows with scripting

By the end of the first section, you will have learned how to search the Fusion App Store using various filters for operating systems, app categories, and app costs. You will also learn how to download, install, and run apps to perform certain automation actions. In the next section, you will also gain a good understanding of how to customize Fusion's machine library and create your own machines, with associated print settings and post files. Using this technique, you will be able to create your future additive setups much more efficiently. In this section, you will also learn how to save your frequently used inputs as user defaults for various operations, such as part orientation, part arrangement, and support structure generation. You will also learn how to create templates that can contain multiple actions with your custom inputs, and how to reuse those templates in subsequent additive setups. By the end of the chapter, you will know how to create and customize your first Python-based script to automate the entire additive workflow for an FFF machine, with one or multiple components.

Technical requirements

All topics covered in this chapter are accessible to Personal, Trial, Commercial, Startup, and Educational Fusion 360 license types. If you intend to create or customize scripts with the Python programming language to be able to automate your additive setups in Fusion 360, you will also need to install the Visual Studio Code software from Microsoft: `https://code.visualstudio.com/`.

How to download and install this software will be highlighted within the relevant section of this chapter.

The lesson files for this chapter can be found here: `https://github.com/PacktPublishing/3D-Printing-with-Fusion-360`

Using apps from the Autodesk App Store

Fusion 360 has a lot of built-in functionality around design and manufacturing using 3D printing. However, there are certain instances where you may benefit from additional functionality to automate your workflow. In such cases, you may want to take a look at the Autodesk App Store to see whether there are any applications readily available for you to utilize, in order to complete a certain task without having to manually execute multiple steps.

You can access the Autodesk App Store in a web browser by navigating to `https://Apps.autodesk.com/FUSION/en/Home/Index`. You can also access the **Autodesk App Store** command located within the **ADD-INS** panel of the **UTILITIES** tab in most workspaces, including the **DESIGN** workspace, as shown by callout **1** in *Figure 14.1* .

Once the Autodesk App Store is visible in your web browser, you will notice that the product is set to Fusion 360 in the top-left corner of the web page. Using this App Store, you can look through the featured apps or search for a specific app by typing the name of the app in the search bar. You can also filter apps related to **3D Printing** by selecting the appropriate filter on the left-hand side, as shown with callout **2**.

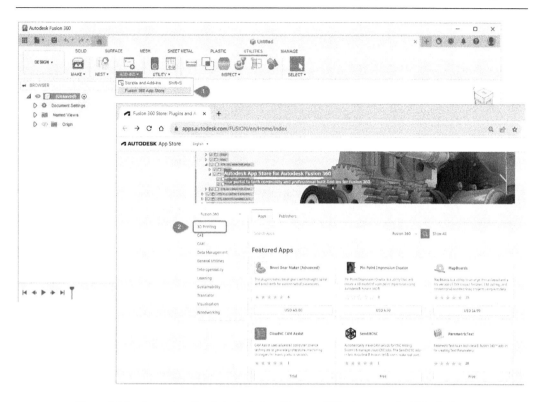

Figure 14.1 – Accessing the Autodesk App Store and filtering apps for 3D printing

After selecting **3D Printing** as a filter, the App Store displays a more advanced search and filter toolset, so you can look for the most relevant, the newest, or the most downloaded apps using the appropriate option from the title bar, as shown by callout **1** in *Figure 14.2*. The Autodesk App Store includes numerous applications that are free to download and use. However, there are certain apps that require a one-time fee or a subscription. Using the filters, you can also rank apps based on rating and price. Using the browser on the left side of the App Store, as shown with callout **2** in the same figure, you can filter free apps, apps that offer a trial option, or paid apps. You can also filter apps based on the operating system, such as **Windows** or **Mac OS**.

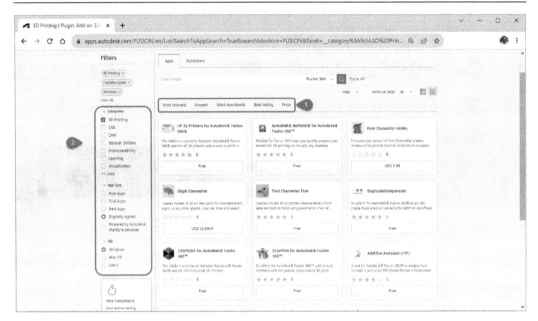

Figure 14.2 – Autodesk App Store with various filtering and ranking mechanisms

In this section, we will demonstrate how to download and utilize a free app from the Autodesk App Store, named 3D Printing Essentials. This app was developed by Autodesk and is available for both macOS and Windows 64-bit operating systems. After navigating to the app, using either the filters we highlighted previously or by searching its name using the search bar, you will see the details of the app, as shown in *Figure 14.3*. In order to download and install this app, you first have to log in to the Autodesk App Store by selecting the **Sign In** option, as shown in callout **1**. Then, you will have to select the operating system, as shown in callout **2**, so that you download the correct version of the app for your computer. After that, you will be able to select the **Download** option, as shown in callout **3**, to download the installer.

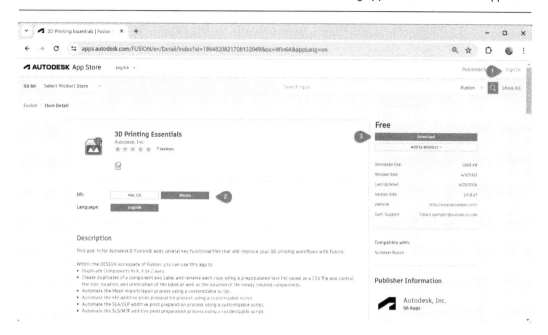

Figure 14.3 – Signing in to the App Store to download an app for Fusion 360

Clicking the **Download** button will save a local copy of the installer on your hard disk. In this example, we will download the **Win64** version of the app. As you can see by callout **1** in *Figure 14.4*, the 3DPrintingEssentials.msi file has been saved to the Downloads folder. After navigating to the appropriate folder, we can open the file, as shown in callout **2**. This will start the installation, as shown in callout **3**.

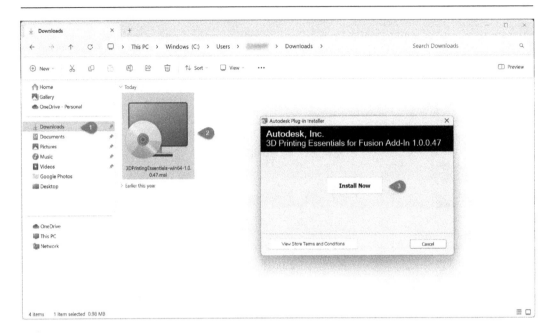

Figure 14.4 – Downloading the MSI file and starting the installation for the app

Once the add-in is installed, we will need to make sure that it is running. In order to start or stop an add-in we installed, we will have to go back to the **Scripts and Add-Ins** dialog within Fusion 360, which is shown in *Figure 14.5*. Then, you will need to go to the **Add-Ins** tab of the **Scripts and Add-Ins** dialog and select the newly installed 3DPrintingEssentials add-in under the **My Add-Ins** folder, as highlighted by callout **2**.

If you see a circular pattern of dots to the right of a given add-in, it means that the add-in is already running. If all you see is the name of the add-in, with no icon to the right of it, it means that the app has been stopped.

You can select a given add-in and run/stop it by selecting the appropriate command at the bottom of this dialog. If you wish, you can also select the checkbox at the bottom of the dialog to run a given add-in on startup, as indicated by callout **3** in the following figure.

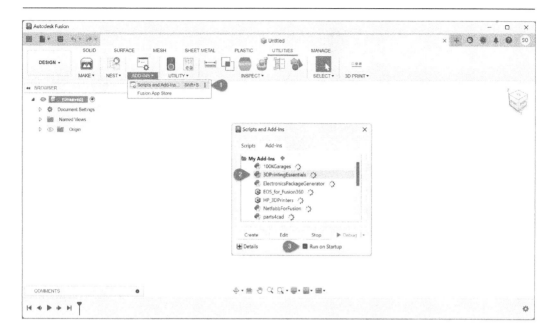

Figure 14.5 – Running add-ins at the time of startup of Fusion 360

Now that our app is installed and running, we are ready to use it with an our workflow. This add-in includes a new set of commands within the **3D PRINT** panel, under the **SOLID** tab of the **DESIGN** workspace, as shown in *Figure 14.6*. If you hover your mouse over the new command, named **Label**, you will see a new icon appear to the right of it. This vertical ellipsis icon is highlighted with callout **1** in the following figure. After selecting this icon, you will be able to activate the **Pin to Toolbar** checkbox, as indicated by callout **2**. This action will add the **Label** icon to the ribbon within the **3D PRINT** panel, as indicated by callout **3**.

If you select the **Duplicate Components** command from either the **3D PRINT** panel, a new dialog called **DUPLICATE COMPONENTS** will appear, as shown in *Figure 14.6*. This dialog will allow you to select a component and create multiple copies of the same component quickly. You have the ability to control the number of copies and the placement of the duplicated components. You can choose to keep the copies at the same location or arrange them in the *X*, *Y*, and *Z* axes, with a specific gap between the bounding boxes of each component.

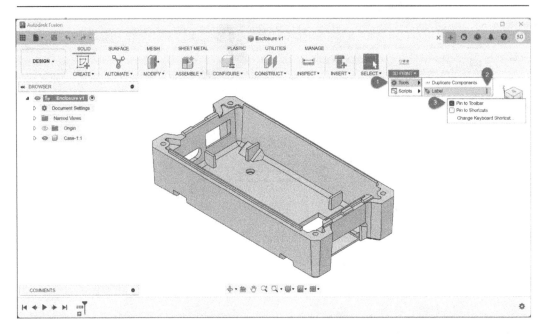

Figure 14.6 – Locating the Duplicate Components and Label command within the 3D PRINT panel

In this example, we have chosen to create seven additional copies of the original component by placing 2 parts in X , Y and Z axes each. This action creates 7 additional components for a total of 8 components, as shown in *Figure 14.7*. Each copy-and-paste action is also documented with a feature in the timeline.

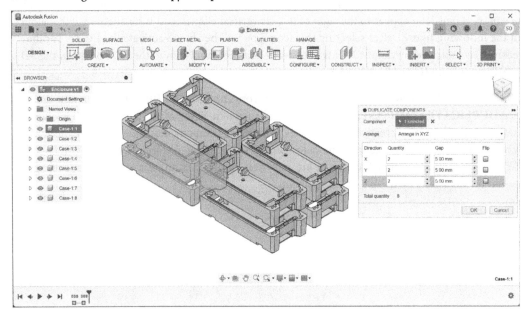

Figure 14.7 – Duplicating a single component to create a 8-component assembly

Fusion 360 has native capabilities to copy and paste components. However, if you wanted to create multiple components with the same location as the original component, you would have to utilize the copy-and-paste commands in sequence as many times as the number of components you wish to create. This could take a long time, depending on the number of copies you wish to make. Utilizing an add-in, such as the one we demonstrated in this section, which automates this process, will save you a significant amount of time and eliminate any potential errors you could introduce during the copy-and-paste process, as the number of components increases.

In this section, we mainly highlighted how to automate our workflows within the **DESIGN** workspace using add-ins. In the next section, we will focus on how to automate certain workflows within the **MANUFACTURE** workspace by customizing our libraries for machines, print settings, and presets.

Customizing presets and using templates

In previous chapters, we demonstrated how to create an additive setup using Fusion 360's **MANUFACTURE** workspace. We also showcased how to orient parts, arrange them within the build volume, and generate supports. We highlighted how to simulate the additive toolpath and generate the necessary machine file.

In this section, we will showcase how to customize Fusion's machine and print setting library in order to speed up the process of creating an additive setup. We will also touch on how to customize the default inputs for various dialogs within an additive setup, and how to create presets to combine multiple custom support operations into a single command, effectively generating support structures for our models.

Even though the process of creating an additive setup is straightforward, it does involve selecting a machine and a print setting explicitly for every single setup. Within the **MANUFACTURE** workspace of Fusion 360, we can access the **Machine Library** dialog within the **MANAGE** panel, as shown in callout **1** of *Figure 14.8*. We can also copy and paste 3D printers from the Fusion 360 library to our own cloud, local, or linked libraries as shown in callout **2**. The following figure shows a 3D printer named **Prusa i3 MK3S+**, saved within the **Local** folder under `My machines`. Whenever we copy and paste a machine from the Fusion 360 library of machines to a folder within `My machines`, the associated post file (`*.cps`), as highlighted with callout **3**, is automatically assigned. However, we still have to designate a print setting when creating an additive setup with this machine. In order to expedite the setup creation process, we can also connect a print setting with this machine by clicking the folder icon shown in callout **4**. After selecting **PLA (Direct Drive).printsetting** from the print setting library, our local machine is now ready to be utilized in the setup creation. After selecting this machine from the local folder of `My machines` during the setup creation, we no longer have to select a print setting, as this print setting is automatically associated with the **PLA direct drive** print setting within the machine library.

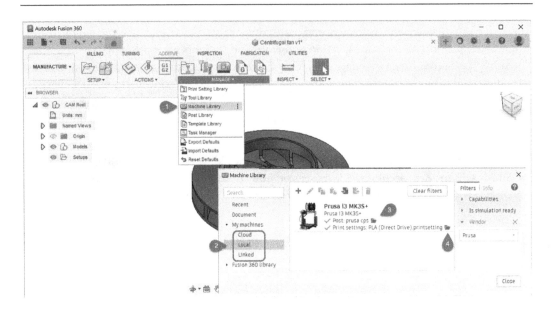

Figure 14.8 – Associating a print setting to customize a machine

After creating an additive setup using this machine and print setting, we may need to orient our components within the build volume of the 3D printer. *Figure 14.9* shows how you can access the **Automatic Orientation** command within the **POSITION** panel of the **ADDITIVE** tab. In this example, we can select our component, named Centrifugal fan, as the target component to be oriented.

> **Important note**
>
> As covered in detail in *Chapter 8*, when calculating the support area and support volume, the automatic orientation command looks for segments of a part that has an overhang angle below a certain angle from the horizontal plane, in order to rank the outcomes of the orientation study.

Any input within a dialog in a manufacturing setup can be customized by selecting the vertical ellipsis icon to the right of the input field, as shown in *Figure 14.9*. After selecting the ⋮ icon, a drop-down menu will be displayed, which will allow you to save your current input as the user default. By customizing an input and saving it as the user default, you can reuse it the next time you execute any of the positioning or support actions within an additive setup.

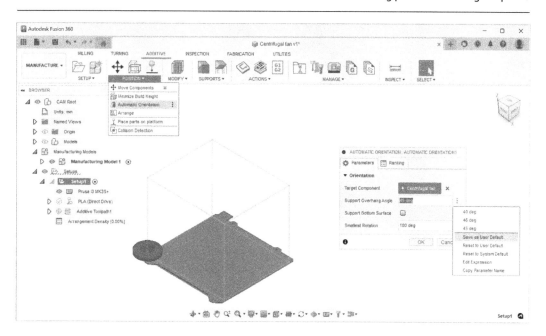

Figure 14.9 – Saving a custom input as the user default

Another useful option within the same menu is the ability to copy the parameter name of any input field. *Figure 14.10* shows the **General** tab of the **SOLID VOLUME SUPPORT** dialog. As a reminder, you can access this dialog by selecting the **Solid Volume Support** command from the **SUPPORTS** panel of the **ADDITIVE** tab, for an additive setup with an FFF 3D printer.

As you can see in the following figure, this tab has two input fields – **Top Distance to Part** and **Bottom Distance to Part**. The default inputs for these fields are 0 . 3 millimeters. Each input field has a parameter name associated with it, which you can reuse within any dialog in the **MANUFACTURE** workspace and while creating scripts or add-ins using Fusion's API. These parameter names can also be utilized as a part of an equation either within the **Edit Expression** dialog or while entering inputs into other input fields.

Figure 14.10 shows how we can simply copy the parameter name, as shown in callout **1**, and paste it into a different input field, as shown in callout **2**, to create an equation where are the **Top Distance to Part** input is half the value of the **Bottom Distance to Part** input.

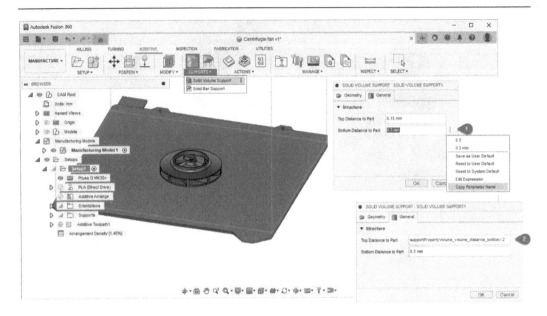

Figure 14.10 – Editing inputs using parameter names

Now that we have covered how to customize user inputs and utilize parameters, we can switch gears and focus on combining multiple operations, such as arranging parts, performing automatic orientation studies, and creating support structures in a single template.

To demonstrate how we can utilize templates, let's create a new Fusion document named Centrifugal fan-SLA. This Fusion document has an additive setup with an SLA printer, as shown in *Figure 14.11*. The printer we are using for the setup is a **Prusa SL1S SPEED** printer, and the print setting is set to **Prusa SL1S SPEED - Prusa Orange Tough - 0.1mm Fast**. The part is already arranged and oriented within the build volume. As you can see from *Figure 14.11*, we have already created a down-oriented point bar support for this component, and we will create a second support using the **Bar Support** action located within the **SUPPORTS** panel of the **ADDITIVE** tab. During the bar support creation, we change the input field for **Bars per Group** to **4**, enable the checkbox named **Brace**, and select the **K** brace type.

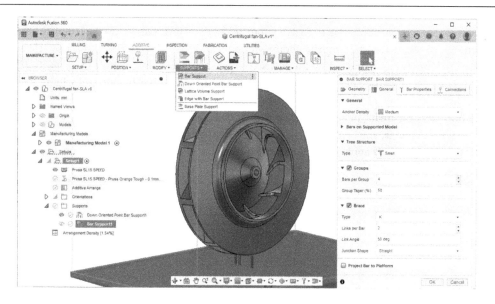

Figure 14.11 – Adding a bar support to a part within an additive setup with an SLA printer

After creating both of these support structures, we can select them from the browser. If you right-click on the selected support actions, Fusion will display a new menu, as shown in *Figure 14.12*. Within this menu, we can select **Store as Template**, which will bring up the **Store as template** dialog. In this dialog, we can name our template and type a custom description. We can also choose whether to store this template in our **Local** template library or the **Cloud** template library.

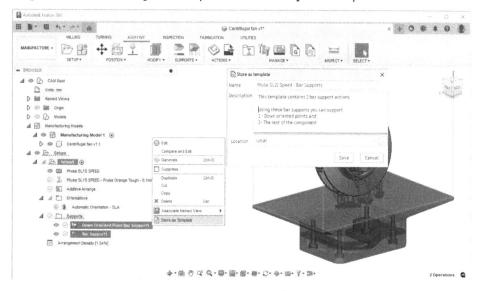

Figure 14.12 – Saving two support actions as a template

After we store our custom support actions as a template, they will be listed within the template library of Fusion 360. We can access the template library within the **MANUFACTURE** workspace by selecting the **Template Library** command within the **MANAGE** panel, as shown by callout **1** in *Figure 14.13*. If you navigate to the Local folder within **My templates**, as shown with callout **2**, you will be able to visualize the newly created template named **Prusa SL1S Speed - Bar Supports**, as shown by callout **3**. In the future, if you create an additive setup with an SLA printer and want to utilize this template with two support actions, you can simply select the **Create operations** icon, as shown with callout **4** in the following figure, and apply it to your additive setup.

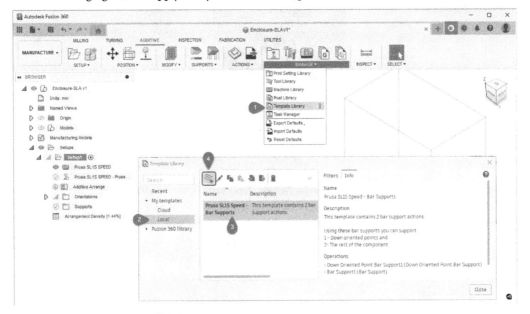

Figure 14.13 – Managing templates and adding them to existing setups

Figure 14.14 shows a new Fusion design document named Enclosure-SLA. This document already contains an additive setup with the same printer and print setting. The part has been oriented and placed within the build volume. Two support actions have also been added using the template library, as shown in *Figure 14.13*. Once you add support actions from a template to a setup, you will notice that the support actions have errors, as indicated by the white exclamation mark within a red diamond icon, alerting you that they are missing a reference. By selecting the individual support actions, you can right-click and choose the **Edit** command, as shown by callout **1** in *Figure 14.14*. Then, you can select the model you wish to support from the canvas, as indicated with callout **2**, and select **OK** to generate the support, as shown by callout **3**.

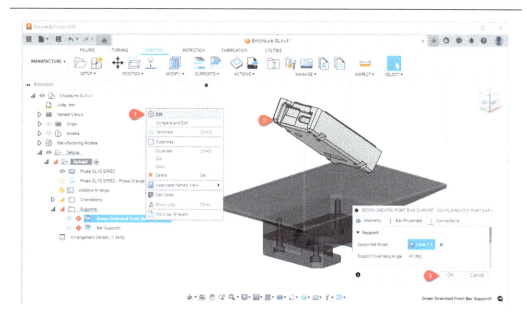

Figure 14.14 – Editing a support action to select a missing reference(s)

After defining the model to support each support action within the template, and generating the support structures, Fusion 360 will display the supports. The outcome of reusing the two support action templates we created while setting up the Fusion document named `Centrifugal fan-SLA`, in the document named `Enclosure-SLA`, is shown in *Figure 14.15*.

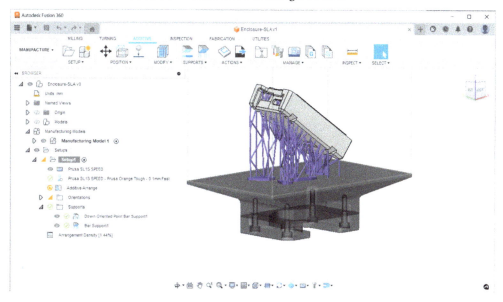

Figure 14.15 – Generating supports after selecting the model to support

Customizing machines within the machine library and selecting an associated print setting for your printer will speed up the creation of an additive setup. Customizing the user defaults will allow you to utilize the automatic orientation and additive arrange functionality with your preferences. Creating custom templates for orientation, arrangement, or support structures will enable you to create consistent additive setups, based on your machine's print settings and materials, so that you eliminate any inconsistencies that may be introduced due to human error. In the next section, we will highlight how to automate the entire process using scripts so that we can completely eliminate any interaction with Fusion 360's **MANUFACTURE** workspace, in order to create an additive setup.

Automating additive workflows with scripting

In the previous section, we covered how to save our preferences as user defaults and create templates while generating an additive setup, in order to reapply those templates to subsequent additive setups.

Using these techniques will minimize our interaction with the user interface of Fusion 360 and help us eliminate inconsistencies between each setup. We can take this concept to the next level by utilizing scripts, eliminiating our interaction with Fusion 360's **MANUFACTURE** workspace entirely.

If we consistently use the same type of machines, print settings, part orientation and arrangement options, and support structure settings, we can benefit from utilizing a custom script to automate the entire workflow. As mentioned in the first section of this chapter, we can access the **Scripts and Add-Ins** command within the **ADD-INS** panel of the **UTILITIES** tab, as shown by callout **1** in *Figure 14.16*. The **Scripts** tab within the **Scripts and Add-Ins** dialog lists several Python and C++ based sample scripts that you can execute to automate certain workflows, as indicated by callout **2**. After selecting a certain script, you can expand the **Details** section, as indicated by callout **3**, to read more about what that script does and inspect its installation path.

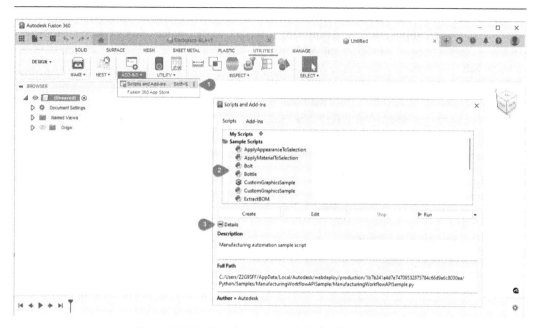

Figure 14.16 – Sample scripts included with Fusion 360

If you want to create your own Python script or edit an existing one, you will need to download Visual Studio Code in order to edit or debug a given script. If you do not have Visual Studio Code software installed on your computer, after selecting a script and pressing **Edit**, Fusion will display the **Download Visual Studio Code** dialog, as shown in *Figure 14.17*. If you select **OK**, it will automatically download and install the Visual Studio Code software from Microsoft.

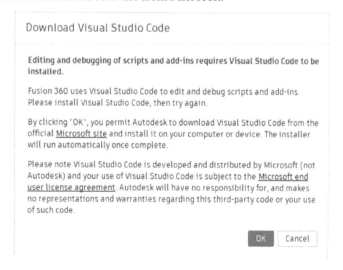

Figure 14.17 – Downloading and installing Visual Studio Code to edit and debug Python-based scripts

Fusion 360 has a rich **application programming interface** (**API**), which allows you to customize software by creating your own scripts and add-ons. The apps we covered in the first section of this chapter, which we downloaded from the Autodesk App Store, were all created using Fusion's API to automate certain workflows. If you find yourself in a position where you have an idea of how to automate a certain workflow but an app does not exist for it, you can create your own scripts or apps. To learn more about scripting for Fusion 360, refer to the Fusion 360 API documentation at the following link: `https://help.autodesk.com/view/Fusion360/ENU/?guid=GUID-A92A4B10-3781-4925-94C6-47DA85A4F65A`. This link has the complete product documentation for the Fusion API and is shown in *Figure 14.18*:

Figure 14.18 – Online documentation for the Fusion 360 API

Within this web page, if you scroll down and navigate to the **Sample Programs** section, as indicated by callout **1** of *Figure 14.19*, you will be able to locate the **CAM** section. After expanding the **CAM** heading, as shown by callout **2**, you will find **Additive Manufacturing API Sample** within the **General** folder, as shown by callout **3**. This page includes a Python-based code sample, which automatically creates an additive setup with an FFF printer and an appropriate print setting. In order to use this code sample, you will first need to copy the Python code. Select the **Copy Code** command, as indicated by callout **4** in *Figure 14.19*.

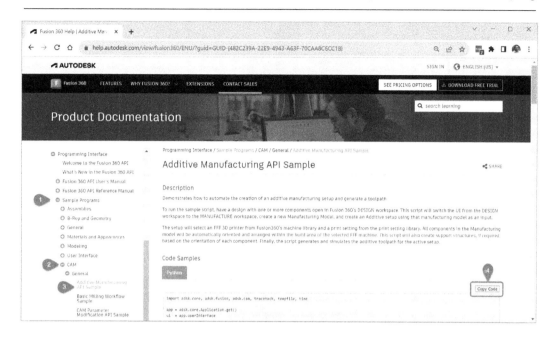

Figure 14.19 – Locating the additive manufacturing API sample

Within the **Scripts and Add-Ins** dialog of Fusion, you can select the + icon next to **My Scripts** to create a new script. This action will populate the **Create New Script or Add-In** dialog, as shown in *Figure 14.20*. In this example, we will create a Python-based script, so we have to select the radio buttons for **Script** and **Python**. We will have to give our script a new name; let's use `FFF Setup`. Then, we can enter a description. We can also enter our name into the **Author** field and define the folder location where the Python script file will be saved. After entering all the necessary information, click the **Create** button to generate an empty Python script.

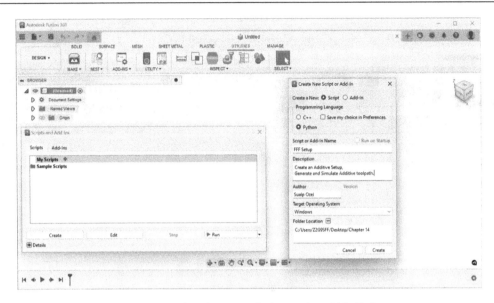

Figure 14.20 – Creating a new Python script within Fusion

Once we have a new script created, it will be listed within the **Scripts and Add-Ins** dialog under the **My Scripts** folder. *Figure 14.21* shows a new script with a Python icon, followed by FFF Setup, as indicated with callout **1**. After selecting this script, press **Edit**, as shown in callout **2**. This action will launch the Visual Studio Code software and allow us to edit the script. We can select the existing text within the script and replace it, by pasting the Additive Manufacturing API sample code we copied from the Fusion 360 online documentation.

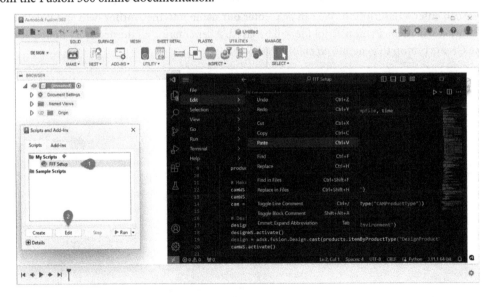

Figure 14.21 – Editing the Python script

As explained in the description section on the `help.autodesk.com` page, this script creates an additive setup, with all the components using an FFF printer and print setting. Then, it automatically orients and arranges all the parts on the build plate. It also adds volume supports to all the parts that require support structures, based on the critical overhang angle defined within the script. Finally, it generates the additive toolpath and simulates it. Technically, such a script can also be further expanded to generate the G-code. However, it is always a good idea to visually observe the setup and the generated toolpath before generating the machine file and starting a print.

All the actions this script takes are explicitly defined in the code, and they can be customized based on your specific printer, print settings, and additive manufacturing technology. In *Figure 14.22*, you can see the line items related to the print settings and the machine model that the script utilizes. You can type in a different machine model or a different print setting by entering them within quotation marks, highlighted by callout **1** and callout **2**. After saving the script and executing it, Fusion will create an additive setup for your printer and print settings.

Figure 14.22 – Editing a script to customize it for your printer and print settings

Now that we have highlighted how to create a script and customize it, let's demonstrate how to use it with a new Fusion design document. To showcase what this script does, we will use a Fusion document named `Front Bumper Bracket`. After opening this Fusion document, we can use the *Shift + F* shortcut to access the **Scripts and Add-Ins** dialog. After selecting our custom script, named `FFF Setup`, we can select **Run** to execute it, as shown in *Figure 14.23*.

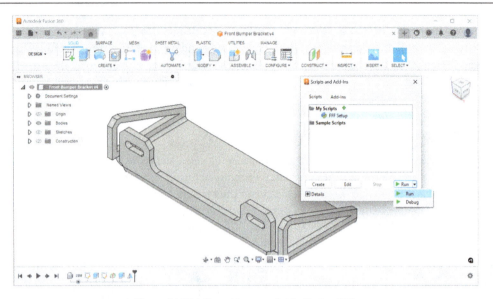

Figure 14.23 – Executing a script in Fusion 360

After a few seconds, you will notice that the active workspace has changed from **DESIGN** to **MANUFACTURE**, as shown in *Figure 14.24*. This script created a new manufacturing model called My Manufacturing Model - FFF. It also created a new additive setup with the **Prusa i3 MK3S+** printer and the **PLA (Direct Drive)** print setting. It automatically oriented the component so that it requires the least amount of support structures. It then arranged the component so that it is in the center of the build plate. It also added the necessary solid volume support and created the additive toolpath. The script left us with a simulated additive toolpath visualization, as shown in *Figure 14.24*.

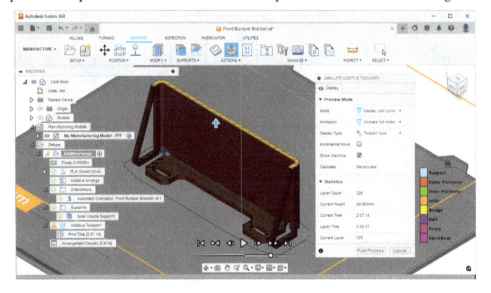

Figure 14.24 – The result of executing the FFF setup script

As mentioned before, all of these actions are things we could have done manually using Fusion's various commands and dialogs. However, by utilizing a script, we eliminated the need for manual interaction with the software. The time saved by automating a single component may not be significant, but imagine if you had 10 unique components you needed to 3D print. Going through these steps while interacting with numerous dialogs for each component would be very time-consuming. By automating the entire workflow with a single script, we not only eliminate errors that a user may introduce by mistyping an input but also save time in the print preparation steps. Using automation capabilities with API-driven scripts, you can minimize errors and improve your productivity.

Summary

In this chapter, we talked about how to automate our workflows when creating additive setups and generating toolpaths. We started the chapter by highlighting how to use the Autodesk App Store to search for apps that can improve our productivity for certain design and manufacturing workflows. We demonstrated how to download, install, and use an app named `3DPrintingEssentials`, in order to quickly create multiple copies of the same component so that we can print multiple parts within the same build volume.

Then, we learned how to associate print settings with a given machine so that we can select the machine and print setting combination with ease when creating an additive setup. We then highlighted how to customize inputs for various dialogs, such as part orientation or additive arrange, and save our inputs as user defaults so that the next time we use those commands, we don't have to edit the inputs to match our preferences. We also learned how to inquire the parameter name for each input field, and how to use those parameter names as inputs in subsequent input fields. We ended the chapter by learning about creating multiple support actions for a given setup and saving those actions as a template. We highlighted where templates are saved within the template library and how to reuse those templates in subsequent additive setups, saving time and minimizing user input errors.

We ended the chapter by demonstrating the use of scripting with programming languages, such as Python, in order to automate additive workflows within Fusion 360. We showcased how to create scripts and edit them using Visual Studio Code, addressing our automation needs. We created an example Python script and showed how it can be used to automate the entire additive workflow, from setup creation to part orientation, part placement, support generation, and toolpath creation.

Using all the information we have focused on in this chapter, you can save time by automating your repetitive tasks and eliminate any potential errors you may introduce if you were to create your additive setups manually.

Index

www.packtpub.com

Subscribe to our online digital library for full access to over 7,000 books and videos, as well as industry leading tools to help you plan your personal development and advance your career. For more information, please visit our website.

Why subscribe?

- Spend less time learning and more time coding with practical eBooks and Videos from over 4,000 industry professionals

- Improve your learning with Skill Plans built especially for you

- Get a free eBook or video every month

- Fully searchable for easy access to vital information

- Copy and paste, print, and bookmark content

Did you know that Packt offers eBook versions of every book published, with PDF and ePub files available? You can upgrade to the eBook version at packtpub.com and as a print book customer, you are entitled to a discount on the eBook copy. Get in touch with us at customercare@packtpub.com for more details.

At www.packtpub.com, you can also read a collection of free technical articles, sign up for a range of free newsletters, and receive exclusive discounts and offers on Packt books and eBooks.

Other Books You May Enjoy

If you enjoyed this book, you may be interested in these other books by Packt:

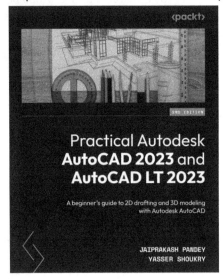

Practical Autodesk AutoCAD 2023 and AutoCAD LT 2023

Jaiprakash Pandey, Yasser Shoukry

ISBN: 978-1-80181-646-5

- Understand CAD fundamentals like functions, navigation, and components
- Create complex 3D objects using primitive shapes and editing tools
- Work with reusable objects like blocks and collaborate using xRef
- Explore advanced features like external references and dynamic blocks
- Discover surface and mesh modeling tools such as Fillet, Trim, and Extend
- Use the paper space layout to create plots for 2D and 3D models
- Convert your 2D drawings into 3D models

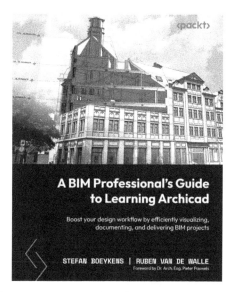

A BIM Professional's Guide to Learning Archicad

Stefan Boeykens, Ruben Van de Walle

ISBN: 978-1-80324-657-4

- Create an architectural model from scratch using Archicad as BIM software
- Leverage a wide variety of tools and views to fully develop a project
- Achieve efficient project organization and modeling for professional results with increased productivity
- Fully document a project, including various 2D and 3D documents and construction details
- Professionalize your BIM workflow with advanced insight and the use of expert tips and tricks
- Unlock the geometric and non-geometric information in your models by adding properties and creating schedules to prepare for a bill of quantities

Packt is searching for authors like you

If you're interested in becoming an author for Packt, please visit `authors.packtpub.com` and apply today. We have worked with thousands of developers and tech professionals, just like you, to help them share their insight with the global tech community. You can make a general application, apply for a specific hot topic that we are recruiting an author for, or submit your own idea.

Share Your Thoughts

Now you've finished *3D Printing with Fusion 360*, we'd love to hear your thoughts! Scan the QR code below to go straight to the Amazon review page for this book and share your feedback or leave a review on the site that you purchased it from.

https://packt.link/r/1-803-24664-2

Your review is important to us and the tech community and will help us make sure we're delivering excellent quality content.

Download a free PDF copy of this book

Thanks for purchasing this book!

Do you like to read on the go but are unable to carry your print books everywhere? Is your eBook purchase not compatible with the device of your choice?

Don't worry, now with every Packt book you get a DRM-free PDF version of that book at no cost.

Read anywhere, any place, on any device. Search, copy, and paste code from your favorite technical books directly into your application.

The perks don't stop there, you can get exclusive access to discounts, newsletters, and great free content in your inbox daily

Follow these simple steps to get the benefits:

1. Scan the QR code or visit the link below:

https://packt.link/free-ebook/9781803246642

2. Submit your proof of purchase.
3. That's it! We'll send your free PDF and other benefits to your email directly.

www.ingramcontent.com/pod-product-compliance
Lightning Source LLC
Chambersburg PA
CBHW060647060326
40690CB00020B/4548